Springer Tracts in Modern Physics
Volume 147

Springer
Berlin
Heidelberg
New York
Barcelona
Hong Kong
London
Milan
Paris
Singapore
Tokyo

Springer Tracts in Modern Physics

Springer Tracts in Modern Physics provides comprehensive and critical reviews of topics of current interest in physics. The following fields are emphasized: elementary particle physics, solid-state physics, complex systems, and fundamental astrophysics.
Suitable reviews of other fields can also be accepted. The editors encourage prospective authors to correspond with them in advance of submitting an article. For reviews of topics belonging to the above mentioned fields, they should address the responsible editor, otherwise the managing editor.
See also http://www.springer.de/phys/books/stmp.html

Managing Editor

Gerhard Höhler

Institut für Theoretische Teilchenphysik
Universität Karlsruhe
Postfach 69 80
D-76128 Karlsruhe, Germany
Phone: +49 (7 21) 6 08 33 75
Fax: +49 (7 21) 37 07 26
Email: gerhard.hoehler@physik.uni-karlsruhe.de
http://www-ttp.physik.uni-karlsruhe.de/

Elementary Particle Physics, Editors

Johann H. Kühn

Institut für Theoretische Teilchenphysik
Universität Karlsruhe
Postfach 69 80
D-76128 Karlsruhe, Germany
Phone: +49 (7 21) 6 08 33 72
Fax: +49 (7 21) 37 07 26
Email: johann.kuehn@physik.uni-karlsruhe.de
http://www-ttp.physik.uni-karlsruhe.de/~jk

Thomas Müller

Institut für Experimentelle Kernphysik
Fakultät für Physik
Universität Karlsruhe
Postfach 69 80
D-76128 Karlsruhe, Germany
Phone: +49 (7 21) 6 08 35 24
Fax: +49 (7 21) 6 07 26 21
Email: thomas.muller@physik.uni-karlsruhe.de
http://www-ekp.physik.uni-karlsruhe.de

Roberto Peccei

Department of Physics
University of California, Los Angeles
405 Hilgard Avenue
Los Angeles, CA 90024-1547, USA
Phone: +1 310 825 1042
Fax: +1 310 825 9368
Email: peccei@physics.ucla.edu
http://www.physics.ucla.edu/faculty/ladder/peccei.html

Solid-State Physics, Editor

Peter Wölfle

Institut für Theorie der Kondensierten Materie
Universität Karlsruhe
Postfach 69 80
D-76128 Karlsruhe, Germany
Phone: +49 (7 21) 6 08 35 90
Fax: +49 (7 21) 69 81 50
Email: woelfle@tkm.physik.uni-karlsruhe.de
http://www-tkm.physik.uni-karlsruhe.de

Complex Systems, Editor

Frank Steiner

Abteilung Theoretische Physik
Universität Ulm
Albert-Einstein-Allee 11
D-89069 Ulm, Germany
Phone: +49 (7 31) 5 02 29 10
Fax: +49 (7 31) 5 02 29 24
Email: steiner@physik.uni-ulm.de
http://www.physik.uni-ulm.de/theo/theophys.html

Fundamental Astrophysics, Editor

Joachim Trümper

Max-Planck-Institut für Extraterrestrische Physik
Postfach 16 03
D-85740 Garching, Germany
Phone: +49 (89) 32 99 35 59
Fax: +49 (89) 32 99 35 69
Email: jtrumper@mpe-garching.mpg.de
http://www.mpe-garching.mpg.de/index.html

K.-E. Peiponen E.M. Vartiainen
T. Asakura

Dispersion, Complex Analysis and Optical Spectroscopy

Classical Theory

With 46 Figures

Springer

Dr. Kai-Erik Peiponen
University of Joensuu
Department of Physics
P.O. Box 111
80100 Joensuu, Finland
Email: peiponen@cc.joensuu.fi

Dr. Erik M. Vartiainen
Lappeenranta University of Technology
Department of Electrical Engineering
P.O. Box 20
53851 Lappeenranta, Finland
Email: Erik.Vartiainen@lut.fi

Dr. Toshimitsu Asakura
Hokkai-Gakuen University
Departments of Electronics and
Information Engineering
Minami-26 Nishi-11, Chuo-ku
064 Sapporo, Hokkaido, Japan
Email: asakura@eli.hokkai-s-u.ac.jp

Physics and Astronomy Classification Scheme (PACS):
78.20.-e, 42.65.-k, 02.30.-f, 02.70.-c

ISSN 0081-3869
ISBN 3-540-64522-5 Springer-Verlag Berlin Heidelberg New York

Library of Congress Cataloging-in-Publication Data applied for.

Die Deutsche Bibliothek – CIP Einheitsaufnahme

Peiponen, Kai-Erik: Dispersion complex analysis and optical spectroscopy:classical theory /K.-E. Peiponen;E.M. Vartiainen; T. Asakura.– Berlin; Heidelberg; New York; Barcelona; Hong Kong; London; Milan; Paris; Singapore; Tokyo: Springer, 1999 (Springer tracts in modern physics; Vol. 147)
ISBN 3-540-64522-5

Typesetting:
Data conversion by Satztechnik Katharina Steingraeber, Heidelberg using a Springer TEX macro package
Cover design: *design & production* GmbH, Heidelberg

SPIN: 10675051 56/3144 - 5 4 3 2 1 0 – Printed on acid-free paper

Preface

Optical spectroscopy is one of the basic tools in material sciences. There are many techniques, such as transmission and reflection spectroscopies, that can be considered as well-established techniques in optical spectroscopy, but the analysis of optical data has always been a problematic question. Since the early 20th century, simple dispersion formulae to describe the interaction of a material with light have been available. However, dispersion formulae such as that of Lorentz usually give only a qualitative picture of the interaction of light with a material.

An important step in the formulation of general dispersion relations, with a minimum of assumptions about the system, was the derivation of dispersion relations by Kramers and Kronig. In this derivation they employed results from complex analysis.

In physics, and specifically in optical physics, complex analysis has been found to be a useful mathematical tool for obtaining various physical properties of a medium. In optical spectroscopy the basic principle of causality, symmetry properties and the asymptotic behavior of the optical constants provide the frames for various kinds of optical data analysis by dispersion relations. Furthermore, complex analysis provides us with, in addition to the dispersion relations, sum rules that can be used in testing real data and theoretical models.

In modern optics nonlinear optical phenomena have provided new tools for spectroscopical studies. Therefore data analysis by means of dispersion relations and other methods has become an important question. It is therefore also easy to understand why complex analysis, which is already familiar from linear optics, has been employed in nonlinear optics.

The object of this book is to present complex analysis as a tool for dispersion theory in optical spectroscopy. We consider linear and nonlinear optical processes and deal with dispersion relations, conformal mappings, the maximum entropy method and sum rules for the analysis of various optical spectra related to various material structures. For the sake of simplicity we make use of simple classical models to describe the interaction of light with a material. It is the wish of the authors that this book should provide a standard textbook in which one can find the most typical methods for dispersion analysis that are based on complex analysis.

The authors express their deep gratitude to Professor Pertti Ketolainen for careful reading of the manuscript, Mrs Pia Kujansuu for help in typing the present material and also to Mrs Riitta Honkanen for skilfully drawn figures. We thank Mr Pertti Pääkkönen and Mr Ville Hautamäki for helping us to obtain the correct layout of the text.

K.-E. Peiponen is grateful to the Academy of Finland for financial support.

Joensuu,
Lappeenranta
Sapporo
September 1998

K.-E. Peiponen
E.M. Vartiainen
T. Asakura

Contents

1. Classical Dispersion Theory

The interaction of the electromagnetic field of light with a medium causes polarization of the electric charges in the medium. Owing to the dynamical nature of the electromagnetic field, a time-dependent polarization of the electric charges occurs. The polarization arises from both ionic and electronic motion. However, the interaction of the electromagnetic field with the electron system of the medium is usually dominant in optical spectroscopy. In this chapter we consider classical dispersion formulae that can be applied mainly for qualitative description of the optical properties of media. These formulae have their basis in the models and theories of mechanics and electromagnetism.

1.1 Equation of Motion

In classical theory of linear dispersion we have to understand how the light field propagates in and interacts with the medium. The question of light propagation in a medium is postponed to Sect. 1.4; here we concentrate on the interaction process. We consider the motion of an electron, bound to a heavy nucleus, which can be described using the model of a damped harmonic oscillator. The external electric field of the incident light acts as the driving force, experienced by the charge of the electron as a disturbance, and the light interaction is governed by the laws familiar from electromagnetism. Depending on the medium, we usually can expect that the strength of the internal electric field can vary locally. Therefore the driving force should have a local-field correction for the microscopic electric field as described by local-field models such as those of Lorentz and Onsager [1–3]. Here we assume that, on average, the local field is equal to the external electric field of the light. In addition we use the complex expression for the light field $E = E_0 \exp(-\mathrm{i}\omega t)$, which is assumed to be applied in the x-direction of the Cartesian system. The interaction system is visualized in Fig. 1.1.

The motion of an electron due to interaction can be expressed by the Newtonian law of dynamics as follows:

$$m\frac{\mathrm{d}^2x}{\mathrm{d}t^2} + m\Gamma\frac{\mathrm{d}x}{\mathrm{d}t} + \beta x = -eE_0\mathrm{e}^{-\mathrm{i}\omega t}, \tag{1.1}$$

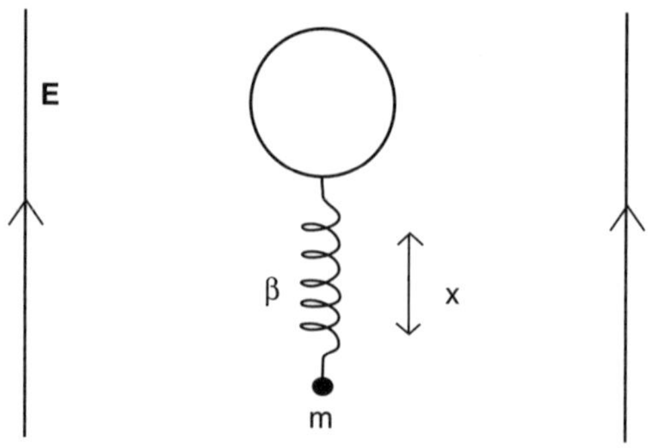

Fig. 1.1. Electric-field-driven oscillator

where m is the mass of the electron, e is the (magnitude of the) electron charge, Γ is the damping parameter (damping can be imagined to be caused by friction that appears between the moving electron and its environment) and β describes the spring coefficient. As usual we re-express the spring coefficient with the aid of the natural angular frequency ω_0 as $\beta = m\omega_0^2$, which describes the electron motion without damping. One straightforward solution of the linear second-order differential equation (1.1) involves a trial function $x(t) = C\exp(-\mathrm{i}\omega t)$, where C is a constant. Substitution of the trial function into (1.1) gives the solution as follows:

$$x(t) = -\frac{eE_0}{m}\frac{\mathrm{e}^{-\mathrm{i}\omega t}}{(\omega_0^2 - \omega^2) - \mathrm{i}\Gamma\omega}. \tag{1.2}$$

The complex solution seems little bit odd since we can expect a real function as a solution. Indeed, it is the complex form of the light field that results also in the complexity of the solution function. The reason why we prefer the use of the complex notation becomes evident when we consider the energy dissipation of the electric field, which is caused by the absorption of radiation in the medium. Then the complex form of (1.2) has a real physical interpretation. Note that in physical reality the angular frequencies and the damping parameter are always real numbers, and therefore the denominator of (1.2) is never zero. Later we deal with complex angular frequency, which permits the denominator to reach the value zero.

If we calculate the modulus (or the amplitude) of $x(t)$, as shown in Fig. 1.2, we obtain a bell-shaped curve, which has its maximum value at a resonant angular frequency given by $\omega_\mathrm{r}^2 = \omega_0^2 - \Gamma^2/2$.

According to classical mechanics, near the resonance frequency very large oscillations can occur, which may even imply damage to the system. In our case, however, where we deal with optical spectroscopy, in the vicinity of the resonant angular frequency the light field is in resonance with the electron oscillators, and the electric field will lose energy to them by absorption, usually without damaging the medium.

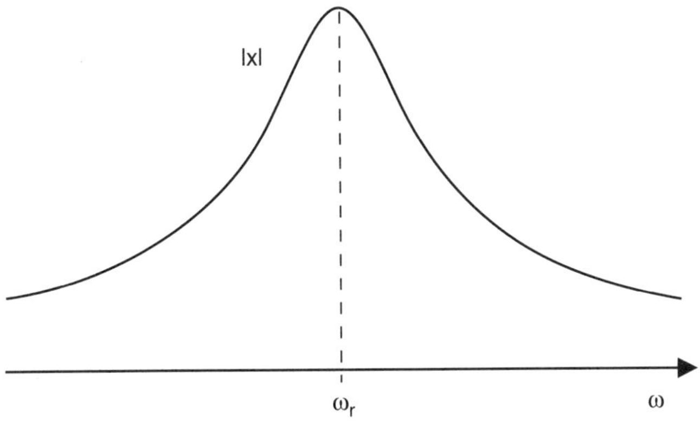

Fig. 1.2. Modulus of complex displacement of oscillator shown in Fig. 1.1

1.2 Maxwell's Equations and Medium Properties

The interaction of light with a medium is characterized by Maxwell's equations. Under the assumption of an isotropic material that obeys Ohm's law for electric conduction and can act in a para- or diamagnetic manner, Maxwell's equations are as follows:

$$\nabla \cdot \boldsymbol{E} = \frac{\rho}{\varepsilon_0}, \qquad \nabla \cdot \boldsymbol{B} = 0,$$

$$\nabla \times \boldsymbol{E} = -\frac{\partial \boldsymbol{B}}{\partial t}, \qquad \nabla \times \boldsymbol{B} = \mu\sigma\boldsymbol{E} + \mu\varepsilon\frac{\partial \boldsymbol{E}}{\partial t}, \tag{1.3}$$

where $\boldsymbol{E}$ is the electric field, $\boldsymbol{B}$ is the magnetic induction, ρ is the charge density other than that due to polarization of the medium, μ is the permeability, σ describes the electric conductivity and ε is the permittivity of the medium. The permittivity, permeability and electric conductivity are parameters that are related to the material properties of the medium. These parameters are usually dependent on the thermodynamical condition of the medium.

Obviously there is a relation between the optical and electrical properties of the medium, since all conductors are nontransparent whereas all transparent materials are insulators. Nevertheless, the transparency of any insulating material is affected also by the presence of grain structure in the material, which can induce partial or total nontransparency. The quality of the insulator can be inspected by several optical techniques, including interferometry [4], speckle patterns [5] and diffractive element sensors [6].

If we consider a perfect transparent insulator then its electric conductivity can be taken to be zero, as well as the external charge density. Under these assumptions Maxwell's equations are reduced to the forms

$$\nabla \cdot \boldsymbol{E} = 0, \qquad \nabla \cdot \boldsymbol{B} = 0,$$

$$\nabla \times \boldsymbol{E} = -\frac{\partial \boldsymbol{B}}{\partial t}, \qquad \nabla \times \boldsymbol{B} = \mu\varepsilon \frac{\partial \boldsymbol{E}}{\partial t}, \tag{1.4}$$

where the linear relations between the electric induction $\boldsymbol{D}$ and electric field $\boldsymbol{E}$, and the magnetic induction $\boldsymbol{B}$ and magnetic field $\boldsymbol{H}$ are as follows:

$$\boldsymbol{D} = \varepsilon \boldsymbol{E},$$

$$\boldsymbol{B} = \mu \boldsymbol{H}. \tag{1.5}$$

In the context of insulators the permeability takes the value of that of a vacuum (μ_0). Linear optics deals with electric charge polarization $\boldsymbol{P}$ that can be expressed with the aid of the linear susceptibility according to the well-known relation

$$\boldsymbol{P} = \varepsilon_0 \chi^{(1)} \boldsymbol{E}, \tag{1.6}$$

where $\chi^{(1)}$ is the so-called linear susceptibility that describes the material properties and ε_0 is the permittivity of vacuum.

The permittivity and linear susceptibility are connected by the relation

$$\varepsilon = \varepsilon_0 \left(1 + \chi^{(1)}\right). \tag{1.7}$$

Later we will generalize the interaction of light with a material so that the polarization has a nonlinear dependence upon the electric field. Nonlinear optical processes occur when the intensity of the incident light is relatively strong, as in the case of a laser beam.

1.3 Lorentz and Drude Models for Linear Susceptibility

There are classical dispersion formulae to describe the linear susceptibilities of insulators and metals. The model for insulators is known as the Lorentz model. Once we know the Lorentz model we can apply it to find an expression for the linear susceptibility. A qualitative description of the linear susceptibility of metals, known as the Drude model, is obtained by a simple inspection of the Lorentz susceptibility. The driven harmonic oscillator, which was considered in Sect. 1.1, is the starting point for the Lorentz dispersion formula. The advantage of the classical models is that we can rather easily relate them to various important material properties.

We begin the derivation of the dispersion formula for the linear susceptibility by considering the induced dipole moment that the electron generates under the effect of the electric field due to the incident light. The dipole moment p is then

$$p = -ex$$

$$= \frac{(e^2 E_0/m)\mathrm{e}^{-\mathrm{i}\omega t}}{\omega_0^2 - \omega^2 - \mathrm{i}\Gamma\omega}, \tag{1.8}$$

where we have substituted the complex solution for x, given in (1.2). According to the theory of electromagnetism, the macroscopic polarization in turn can be expressed with the aid of the number density $\aleph$ of the electrons:

$$P = \aleph p. \tag{1.9}$$

When we insert the result of (1.8) into (1.9) and solve for the linear susceptibility using (1.6), we find a complex linear susceptibility given by the Lorentz form,

$$\chi^{(1)}(\omega) = \frac{\aleph e^2}{m\varepsilon_0} \frac{1}{\omega_0^2 - \omega^2 - \mathrm{i}\Gamma\omega}. \tag{1.10}$$

If we equate the real and imaginary parts we get

$$\mathrm{Re}\left\{\chi^{(1)}(\omega)\right\} = \frac{\aleph e^2}{m\varepsilon_0} \frac{\omega_0^2 - \omega^2}{\left(\omega_0^2 - \omega^2\right)^2 + \Gamma^2\omega^2},$$

$$\mathrm{Im}\left\{\chi^{(1)}(\omega)\right\} = \frac{\aleph e^2}{m\varepsilon_0} \frac{\Gamma\omega}{\left(\omega_0^2 - \omega^2\right)^2 + \Gamma^2\omega^2}. \tag{1.11}$$

The curves of the real and imaginary parts, as a function of the angular frequency, are illustrated in Fig. 1.3.

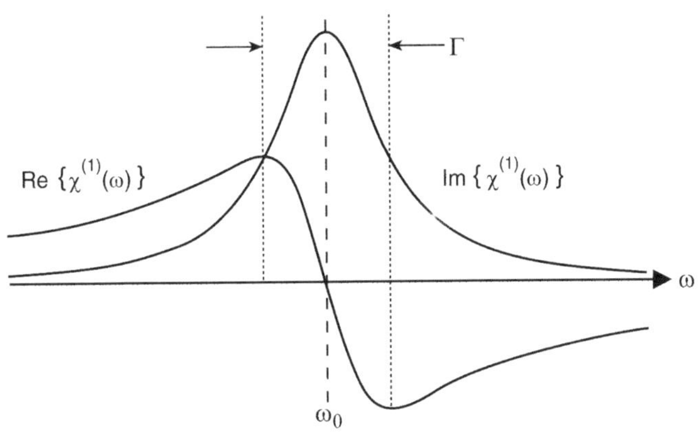

Fig. 1.3. Real and imaginary parts of Lorentz susceptibility

From Fig. 1.3 we observe that the real part has maximum and minimum values. In addition, the real part is monotonically increasing except in a finite angular frequency range between the two extremal points. This region is, for historical reasons, called the region of anomalous dispersion. Actually there is no anomaly at all, but a long time ago the imaginary part was not included in the description of the linear susceptibility, corresponding to letting $\Gamma = 0$ in (1.11). This in turn would mean that the real part of the linear susceptibility would explode when the angular frequency tended to ω_0. Some models, such as those of Cauchy and Sellemeier for the real refractive index of insulators, follow a similar functional behavior to the linear susceptibility under the above assumption. Therefore there was a contradiction or anomaly between the theory and experiments. Indeed, in the early history of optical spectroscopy the theory allowed the possibility that the group velocity of wave packets could exceed the phase velocity and therefore violate the principle of causality (i.e. the response is always at a later instant than the cause that was the origin of the response). This anomaly was not in accordance with the principle of relativity as enunciated by Einstein; the signal velocity could exceed the velocity of light in vacuum. The apparent contradiction was eliminated by the analysis of Sommerfeld and Brillouin [7–9]. The complex nature of the linear susceptibility offers a solution to the problem of achieving consistency between the theory of optical spectroscopy and the principle of relativity.

When we allow the presence of the imaginary part of the linear susceptibility we introduce the possibility of energy loss, which is greatest when the imaginary part of the linear susceptibility has its maximum value, i.e. at the resonant angular frequency, as shown in Fig. 1.3. The power dissipation can be calculated with the aid of the energy density

$$\Psi = \frac{1}{2}\boldsymbol{E}\cdot\boldsymbol{D} = \frac{1}{2}\varepsilon\boldsymbol{E}\cdot\boldsymbol{E}. \tag{1.12}$$

Then the time-average power dissipation is as follows:

$$\begin{aligned}\left\langle\frac{\mathrm{d}\Psi}{\mathrm{d}t}\right\rangle &= \left\langle \mathrm{Re}\left\{E_0\mathrm{e}^{-\mathrm{i}\omega t}\right\}\mathrm{Re}\left\{-\mathrm{i}\omega\varepsilon(\omega)E_0\mathrm{e}^{-\mathrm{i}\omega t}\right\}\right\rangle \\ &= \frac{1}{2}\omega\,\mathrm{Im}\left\{\chi^{(1)}(\omega)\right\}E_0^2,\end{aligned} \tag{1.13}$$

which indicates the role of the imaginary part of the linear susceptibility in the energy loss of the electric field that propagates in the medium.

We have exploited Newtonian mechanics in the derivation of the linear susceptibility of insulators. The Newtonian description implies that as the angular frequency tends to infinity the linear susceptibility is proportional to ω^{-2}. This is a property quite often employed in the development of the dispersion theory in optical spectroscopy.

In insulators the electrons are tightly bound to the nuclei, whereas in the case of metals the conduction electrons, which also interact with light, are free to move. Therefore we can set $\omega_0 = 0$ in the Lorentz model, and then we get for metals the so-called Drude dispersion formula. Note that when the angular frequency tends to zero in the Drude model, the imaginary part of the linear susceptibility tends to infinity. This is consistent with classical electromagnetism, which states that a static (DC) field cannot penetrate into the metal.

So far, we have let the electrons have one resonance frequency. In reality there are usually various other resonance frequencies belonging to different groups of electrons. The above formalism can be generalized with the aid of quantum mechanics. Then the quantized linear susceptibility is a series expansion, where the terms resemble the linear susceptibility of (1.10) but quantum mechanical parameters such as oscillator strength have been introduced. For more details we refer to the book by Wooten [1].

It is important to note the symmetry properties related to the linear susceptibility. Indeed, by inspection of (1.11) we can immediately write

$$\mathrm{Re}\left\{\chi^{(1)}(-\omega)\right\} = \mathrm{Re}\left\{\chi^{(1)}(\omega)\right\},$$
$$\mathrm{Im}\left\{\chi^{(1)}(-\omega)\right\} = -\,\mathrm{Im}\left\{\chi^{(1)}(\omega)\right\}. \tag{1.14}$$

In other words the real part is an even function and the imaginary part is an odd function of the angular frequency. The symmetry properties are quite often referred to in the literature of this field as "crossing relations". The validity of the symmetry relations is not limited to our simple example of the Lorentz oscillator, but they have a general validity. Indeed, according to the theory of linear responses [10], the symmetry can be observed by inspection of the Fourier transform of the polarization [11], $P(\omega) = \varepsilon_0 \chi^{(1)}(\omega) E(\omega)$, in the following manner:

$$P(t) = \varepsilon_0 \int_{-\infty}^{\infty} \chi^{(1)}(\omega) E(\omega) \mathrm{e}^{-\mathrm{i}\omega t} \mathrm{d}t. \tag{1.15}$$

Now because in physical reality the electric field and the polarization are real, we require the symmetry property of the linear susceptibility stated in (1.14), i.e. $\chi^{(1)}(-\omega) = \left[\chi^{(1)}(\omega)\right]^*$, where "*" denotes the complex conjugate.

Although the angular frequency variable, the natural angular frequency and the damping parameter are always real numbers in physical reality, we can replace the angular frequency variable with a complex one, $\widehat{\omega} = \omega + \mathrm{i}v$. Then the linear susceptibility of (1.10) has two singular points, which are called the poles of the function $\chi^{(1)}$. The poles are the roots of

$$\omega_0^2 - \widehat{\omega}^2 - \mathrm{i}\Gamma\widehat{\omega} = 0. \tag{1.16}$$

Since the damping constant is always positive, the poles are always located in the lower half of the complex angular frequency plane. These complex numbers can have both real parts and negative imaginary parts or they can have negative imaginary parts only. In physics the poles are frequently called the resonance points of the system. If we study the properties of the complex linear susceptibility as a function of the complex angular frequency variable, we observe that the linear susceptibility is a holomorphic function [12] in the upper half plane. If the lower half plane is included then the linear susceptibility is holomorphic everywhere except at the poles. Such a function, which is a ratio of two rational functions, is called a meromorphic function [12, 13].

A great deal of dispersion theory in optical spectroscopy depends on the optical constants of the medium being holomorphic. However, it will be pointed out in a later chapter that meromorphic, degenerate nonlinear susceptibilities appear in the context of some nonlinear processes. In such cases the conventional dispersion relations are not valid but have to be replaced with more appropriate relations to make data analysis of optical spectra possible.

The property of holomorphism of the linear susceptibility means that the linear susceptibility is derivable as a function of the complex angular frequency and the derivative is finite. Often the holomorphism is tested with the aid of the Cauchy–Riemann equations [9, 12]. If we restrict the consideration to the upper half of the complex plane, where the linear susceptibility is a well-behaved function, we observe that the Cauchy–Riemann equations are fulfilled as follows:

$$\frac{\partial \operatorname{Re}\left\{\chi^{(1)}(\omega, v)\right\}}{\partial \omega}=\frac{\partial \operatorname{Im}\left\{\chi^{(1)}(\omega, v)\right\}}{\partial v},$$

$$\frac{\partial \operatorname{Re}\left\{\chi^{(1)}(\omega, v)\right\}}{\partial v}=-\frac{\partial \operatorname{Im}\left\{\chi^{(1)}(\omega, v)\right\}}{\partial \omega}. \tag{1.17}$$

The symmetry property of the linear susceptibility, in the complex plane, is now expressed as $\chi^{(1)}(-\widehat{\omega}^*)=\left[\chi^{(1)}(\widehat{\omega})\right]^*$.

We can also present the linear susceptibility in the polar form

$$\chi^{(1)}(\omega)=\left|\chi^{(1)}(\omega)\right| \mathrm{e}^{\mathrm{i} \arg\left\{\chi^{(1)}(\omega)\right\}}. \tag{1.18}$$

The polar form is especially important when we deal with spectra related to the linear reflectance from metals or other nontransparent materials.

1.4 Wave Equation and the Complex Refractive Index

After some vector algebra, using Maxwell's equations, we find the wave equation for both the electric and the magnetic field propagating in an insulator,

$$\nabla^2\{\boldsymbol{E}, \boldsymbol{B}\}=\mu \varepsilon \frac{\partial^2}{\partial t^2}\{\boldsymbol{E}, \boldsymbol{B}\}, \tag{1.19}$$

where electric conductivity is now not allowed. Next we define the complex refractive index N by

$$N^2=\mu_{\mathrm{r}} \varepsilon_{\mathrm{r}}=\left(\frac{\mu}{\mu_0}\right)\left(\frac{\varepsilon}{\varepsilon_0}\right)$$

and

$$N=n+\mathrm{i} k. \tag{1.20}$$

The real part of the complex refractive index, n, is the conventional refractive index whereas the imaginary part, k, which is called the extinction coefficient, describes the attenuation of the electric field in the medium. These two angular-frequency-dependent quantities are often called "optical constants" of the material, although they are not constants in practise but depend, in addition to the angular frequency, also on the external conditions of the material. For insulators the permeability takes the value of that of a vacuum, and the real and imaginary parts of the relative permittivity can be equated to yield

$$\mathrm{Re}\left\{\varepsilon_{\mathrm{r}}(\omega)\right\} = n^2(\omega) - k^2(\omega) = 1 + \mathrm{Re}\left\{\chi^{(1)}(\omega)\right\},$$

$$\mathrm{Im}\left\{\varepsilon_{\mathrm{r}}(\omega)\right\} = 2n(\omega)\,k(\omega) = \mathrm{Im}\left\{\chi^{(1)}(\omega)\right\}. \tag{1.21}$$

The solution of the wave equation is

$$E(x,t) = E_0 \mathrm{e}^{\mathrm{i}((nx/c)-\omega t)} \mathrm{e}^{-k\omega x/c}, \tag{1.22}$$

where c is the light velocity in vacuum, equal to $(\mu_0\varepsilon_0)^{-1/2}$. Equation (1.22) describes a wave that is propagating in the x-direction and is attenuated along the direction of propagation, as shown in Fig. 1.4. In other words, light absorption is present.

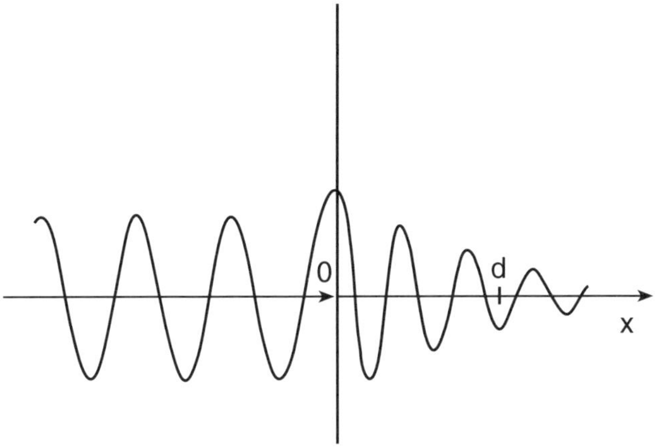

Fig. 1.4. Electric field attenuation in medium. Material is present on the right-hand side of the vertical axis

The light absorption is described by the Beer–Lambert law

$$I = I_0 \mathrm{e}^{-\alpha(\omega)d}, \tag{1.23}$$

where I_0 is the intensity of the incident light, d is the distance and α is the absorption coefficient. The absorption coefficient can be expressed in terms of the extinction coefficient:

$$\alpha(\omega) = \frac{2\omega k(\omega)}{c}. \tag{1.24}$$

In the general case of absorbing materials, including metals, we cannot neglect the conductivity term. However, the above definition of the complex refractive index is generally valid. For bulk metals, in optical spectroscopy we can usually measure the intensity reflectance, which is a function of the optical constants. The optical constants can be considered as intrinsic optical properties of the materials.

We remark that the complex refractive index obeys symmetry relations, just like the linear susceptibility and the permittivity. Furthermore, the complex refractive index can be described as a holomorphic function in the upper half of the complex angular frequency plane. This comes from the fact that if N is holomorphic then N^2 also has to be holomorphic. Since the holomorphic permittivity can be expressed as the square of the complex refractive index, we can imagine that the complex refractive index itself has to be a holomorphic function. In fact, it is a result of complex analysis that a positive integer power of a holomorphic function is also a holomorphic function. This property has far-reaching results, especially when we later deal with sum rules for optical constants.

Finally we derive, with the aid of (1.21), relations between the real and imaginary parts of the complex refractive index and the linear susceptibility, as follows:

$$\begin{aligned}&\left[1+\mathrm{Re}\left\{\chi^{(1)}(\omega)\right\}\right]^2+\left[\mathrm{Im}\left\{\chi^{(1)}(\omega)\right\}\right]^2\\&=n^4(\omega)-2n^2(\omega)k^2(\omega)+k^4(\omega)+4n^2(\omega)k^2(\omega)\\&=\left[n^2(\omega)+k^2(\omega)\right]^2.\end{aligned}\tag{1.25}$$

Taking the square root of both sides of (1.25), and after that either adding or subtracting $1+\mathrm{Re}\left\{\chi^{(1)}\right\}=n^2-k^2$ on both sides of the new equation, we find expressions for n and k:

$$\begin{aligned}n(\omega)=&\left\{\frac{1}{2}\left\{\left[(1+\mathrm{Re}\{\chi^{(1)}(\omega)\})^2+(\mathrm{Im}\{\chi^{(1)}(\omega)\})^2\right]^{1/2}\right.\right.\\&\left.\left.+(1+\mathrm{Re}\{\chi^{(1)}(\omega)\})\right\}\right\}^{1/2},\\k(\omega)=&\left\{\frac{1}{2}\left\{\left[(1+\mathrm{Re}\{\chi^{(1)}(\omega)\})^2+(\mathrm{Im}\{\chi^{(1)}(\omega)\})^2\right]^{1/2}\right.\right.\\&\left.\left.-(1+\mathrm{Re}\{\chi^{(1)}(\omega)\})\right\}\right\}^{1/2}\end{aligned}\tag{1.26}$$

1.5 Complex Reflectivity

The interaction of an electromagnetic field with a medium involves, in addition to absorption and refraction, also reflection of the electric field from the interface between two materials. In the general case both materials can be absorbing media. However, quite often optical spectra are measured so that the light is incident from air. In that case the system is somewhat simpler. As is well known, Fresnel's equations describe the strength of specularly reflected light. Fresnel's equations can be derived by demanding continuity of the tangential components of the electric and magnetic fields at the boundary between the two materials.

We state next Fresnel's equations for the s- and p-polarized-light components (see Fig. 1.5, which shows the polarizations of the electric-field components and the geometry of specular reflection) of the electric-field complex reflectivity, which depends on the optical constants of the medium. The equations are as follows [14, 15]:

$$
\begin{aligned}
r_{\mathrm{s}}(\omega) &= \frac{\cos\theta - \sqrt{N^2(\omega) - \sin^2\theta}}{\cos\theta + \sqrt{N^2(\omega) - \sin^2\theta}} = \frac{\cos\theta - (a + ib)}{\cos\theta + (a + ib)} \\
r_{\mathrm{p}}(\omega) &= \pm\frac{N^2(\omega)\cos\theta - \sqrt{N^2(\omega) - \sin^2\theta}}{N^2(\omega)\cos\theta + \sqrt{N^2(\omega) - \sin^2\theta}} \\
&= \pm\frac{\left(n^2 - k^2\right)\cos\theta - a + \mathrm{i}\left(2nk\cos\theta + b\right)}{\left(n^2 - k^2\right)\cos\theta + a + \mathrm{i}\left(2nk\cos\theta - b\right)},
\end{aligned}
\tag{1.27}
$$

where the + sign denotes the Verdet and the − sign the Fresnel convention [15], and

$$
\begin{aligned}
n^2 - k^2 - \sin^2\theta &= a^2 - b^2, \\
2nk &= 2ab.
\end{aligned}
\tag{1.28}
$$

A practical relation between r_{s} and r_{p} can be found when we set

$$
q = \sqrt{N^2(\omega) - \sin^2\theta}. \tag{1.29}
$$

Then we can deduce that, in the Verdet convention,

$$
\begin{aligned}
r_{\mathrm{s}} &= \frac{\cos\theta - q}{\cos\theta + q}, \\
r_{\mathrm{p}} &= \frac{\left(q^2 + \sin^2\theta\right)\cos\theta - q}{\left(q^2 + \sin^2\theta\right) + q}.
\end{aligned}
\tag{1.30}
$$

From the first equation of (1.30) we can solve for q:

$$
q = \frac{1 - r_{\mathrm{s}}}{1 + r_{\mathrm{s}}}\cos\theta\ ; \tag{1.31}
$$

and then substitute it into the second equation of (1.30). Finally we get a result given by Azzam [16],

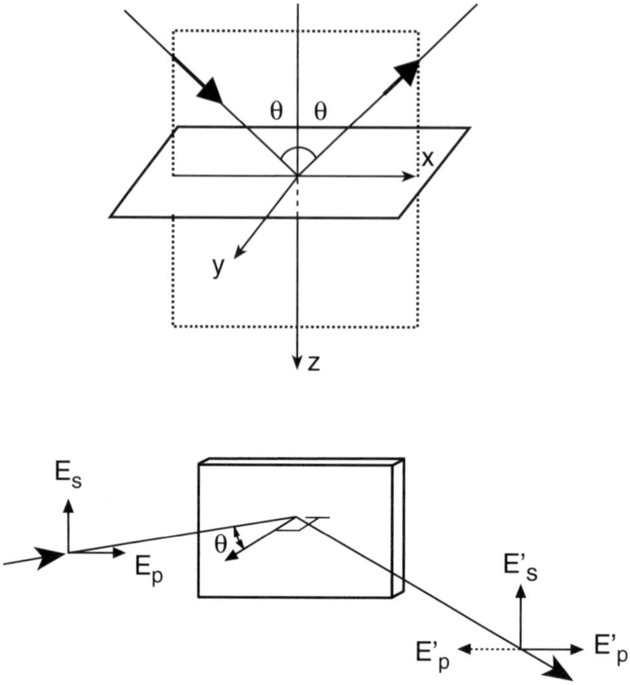

Fig. 1.5. Specular reflection of s- and p-polarized light. In the case of p-polarized light the *solid arrow* for E'_p corresponds to the Fresnel and *dashed arrow* to the Verdet convention

$$r_p = r_s \frac{r_s - \cos 2\theta}{1 - r_s \cos 2\theta}. \tag{1.32}$$

According to (1.32), we can reconstruct r_p whenever the reflectance r_s and the angle of incidence are known. As a special case we obtain the well-known result $r_p = r_s^2$ when $\theta = 45°$. The result of (1.32) is important also because it leads also to a mathematical formula for the ratio r_p/r_s, which can be determined by ellipsometric measurements.

The spectral measurement of reflectance, however, provides us only with information about intensity of the reflected light. Such a signal is theoretically described by the intensity reflectance, which obeys the following Fresnel equations:

$$R_s = r_s r_s^* = \frac{(a - \cos\theta)^2 + b^2}{(a + \cos\theta)^2 + b^2},$$

$$R_p = r_p r_p^* = \frac{\left[(n^2 - k^2)\cos\theta - a\right]^2 + (2nk\cos\theta - b)^2}{\left[(n^2 - k^2)\cos\theta + a\right]^2 + (2nk\cos\theta + b)^2}. \tag{1.33}$$

The complex reflectances can be given in a polar form like that of (1.18). The phase angle information of the complex reflectivity allows us to solve for the optical constants of an opaque material. The phase angle for oblique incidence, for both s- and p-polarized light can be determined. For the case of s-polarized light, let us take the squared modulus on both sides of (1.32), which leads to

$$R_{\mathrm{p}} = |r_{\mathrm{p}}|^2 = R_{\mathrm{s}} \frac{R_{\mathrm{s}} + \cos^2 2\theta - 2R_{\mathrm{s}}^{1/2} \cos 2\theta \cos \varphi_{\mathrm{s}}}{1 + R_{\mathrm{s}} \cos^2 2\theta - 2R_{\mathrm{s}}^{1/2} \cos 2\theta \cos \varphi_{\mathrm{s}}}, \tag{1.34}$$

where φ_{s} is the phase shift of the s-polarized electric field. We can solve the algebraic equation above and get

$$\cos \varphi_{\mathrm{s}} = \frac{\left(R_{\mathrm{s}}^2 - R_{\mathrm{p}}\right) + R_{\mathrm{s}} \left(1 - R_{\mathrm{p}}\right) \cos^2 2\theta}{2R_{\mathrm{s}}^{1/2} \left(R_{\mathrm{s}} - R_{\mathrm{p}}\right) \cos 2\theta}. \tag{1.35}$$

Using (1.32) together with (1.35), we can also solve for the phase angle of p-polarized light. Detailed calculations are given by Azzam [16]. Here we state only the result:

$$\tan\left(\varphi_{\mathrm{p}} - \varphi_{\mathrm{s}}\right) = \frac{R_{\mathrm{s}}^{1/2} \sin^2 2\theta \sin \varphi_{\mathrm{s}}}{R_{\mathrm{s}}^{1/2} \cos \varphi_{\mathrm{s}} \left(1 + \cos^2 2\theta\right) - R_{\mathrm{s}} \cos 2\theta}. \tag{1.36}$$

The calculation of the phase angle for normal-incidence reflectivity is not possible by the above method. Furhermore, in optical spectroscopy we quite often try to measure reflectance at normal incidence. Then the role of the polarization of the incident light is no longer important and the intensity reflectivity for the s- and p-polarized light components becomes the same, namely

$$R = \frac{(a-1)^2 + b^2}{(a+1)^2 + b^2} = \left| \frac{N-1}{N+1} \right|^2 . \tag{1.37}$$

Quite often there is a desire, on one hand, to estimate the strength the of reflectivity when either the real refractive index or the extinction coefficient is known, or, on the other hand, to estimate the real refractive index or the extinction coefficient when the reflectance has been measured. Such estimations may help one to gain more information about the optical properties of, for example, novel materials, especially when the measurement of some of the optical properties is problematic. For this purpose we study the modulus of the reflectivity given in (1.37) in order to find bounds for the reflectivity and the refractive index. We can estimate, using the triangle inequality,

$$1 - \frac{2}{|N+1|} \leq \sqrt{R} \leq 1, \tag{1.38}$$

where the upper bound comes from the principle of energy conservation. In addition, we find the lower bound for the reflectivity

$$R_{\mathrm{LB}}(n) = \left(\frac{n-1}{n+1} \right)^2 \leq R(n,k) . \tag{1.39}$$

The curve of the lower bound of the reflectivity, $R(n)$, is illustrated in Fig. 1.6.

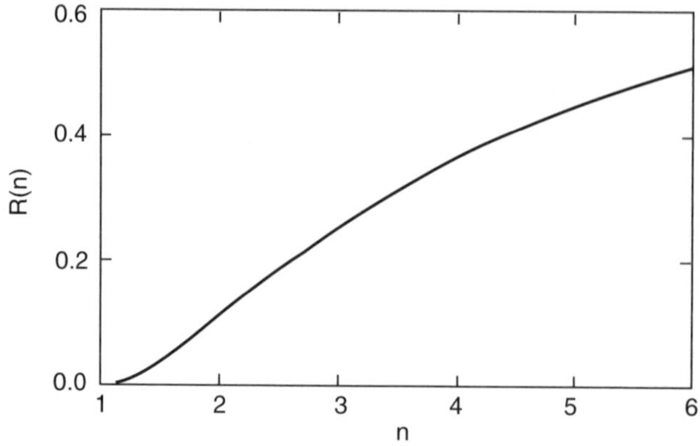

Fig. 1.6. Lower-bound curve for reflectivity

It is also possible to estimate the real refractive index by obtaining from (1.39) its upper bound as follows:

$$n_{\mathrm{UB}}\,(R) \le \frac{1+\sqrt{R}}{1-\sqrt{R}}. \tag{1.40}$$

The corresponding curve for this upper bound $n_{\mathrm{UB}}(R)$ is presented in Fig. 1.7. The estimate of (1.40) is usually relatively good for the case of low reflectance.

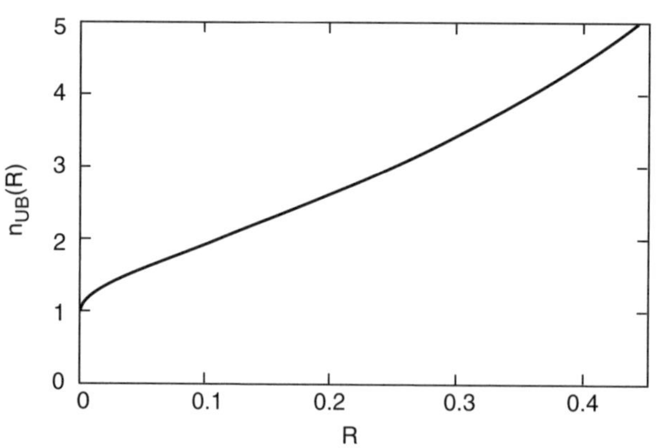

Fig. 1.7. Upper-bound curve for refractive index

The application of Fresnel's equations requires that the interface between the two materials is an ideal, planar surface. Unfortunately, surface roughness, oxidation and contamination of the interface usually have an effect on the strength of the reflectance. This in turn may result in erroneous estimation of the optical constants if they are determined from the reflectance data, e.g. by the Kramers–Kronig relations or by using other techniques devised by Humphreys-Owen [17].

The problem of surface roughness for relatively smooth surfaces can be overcome if the surface statistics are Gaussian. Then, according to Beckmann and Spizzichino [18], the specular reflectivity can be written as

$$R_{\mathrm{ROUGH}}(\omega) = R(\omega)\,\mathrm{e}^{-(2\omega h\cos\theta/c)^2}, \tag{1.41}$$

where h is the optical surface roughness and R is the reflectivity given by Fresnel's equations. Equation (1.41) has importance in industrial metrology, where lasers are employed as light sources to estimate moderate surface roughness of metal surfaces [19]. Beckmann and Spizzichino have also treated the role of diffuse reflection of light in their book. It is worth mentioning that different surface textures can affect the reflectance even if the nominal average surface roughness is the same for two surfaces. This can be observed in Fig. 1.8, where the reflectances of ideal, flat-lapped and ground nickel surfaces used as roughness standards are shown.

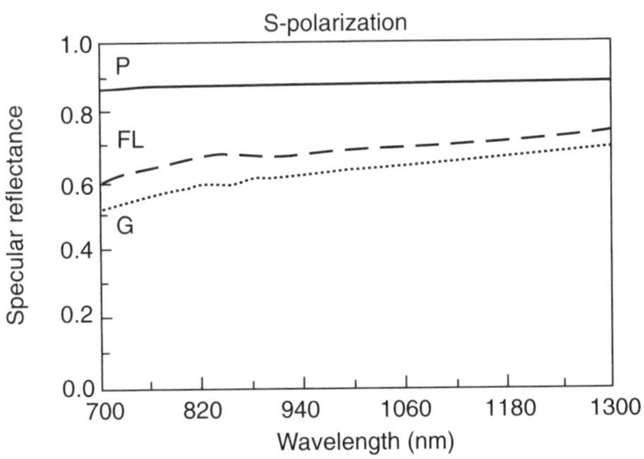

Fig. 1.8. Reflectance (s-polarization) of nickel surfaces: P, perfect; FL, flat-lapped; G, ground. The average surface roughness for the flat-lapped and ground surfaces was 0.1 µm

2. Dispersion Relations in Linear Optics

The optical properties of a medium which are usually obtained directly by spectroscopic measurements are the wavelength-dependent transmittance and reflectance. However, quite often we wish to know the variation with wavelength of the real refractive index in the case of a transmittance measurement, or the optical constants in the case of a reflectance measurement. Unfortunately, the simultaneous measurement of optical constants with conventional spectroscopic devices is not usually possible. An ellipsometer may be considered for that purpose, but usually the spectral measurement range is relatively narrow. At the beginning of this century Kramers and Kronig [20 – 22] derived general relations, now known as the Kramers–Kronig relations, which couple together the angular-frequency-dependent absorption and dispersion. The Kramers–Kronig relations can also be exploited in phase retrieval problems that are related to reflectance data. In that case the optical constants of metals, for example, can be resolved by using an appropriate Kramers–Kronig analysis and Fresnel's equations.

The existence of dispersion relations has a simple but profound physical basis, which is known as the principle of causality. The primitive principle of causality simply states that the response of a system is at a later instant than the reason that caused the response. The principle of causality implies the holomorphicity of the linear susceptibility and optical constants. The logical relations between causality, holomorphism and the existence of dispersion relations have been described by several authors [10, 23, 24].

In this chapter we consider first the principle of causality with the aid of the simple model of the driven harmonic oscillator. Thereafter we derive the Kramers–Kronig relations and describe their application to data inversion from absorption or reflectance data of media.

2.1 Causality

Let us modify the equation of motion (1.1) so that we divide both sides by the mass of the electron and replace the Coulomb driving force due to the light field with a more general time-dependent force. The driving force is denoted by $f(t)$. Now the equation of motion can be given as

$$\frac{\mathrm{d}^2 x}{\mathrm{d}t^2} + \Gamma \frac{\mathrm{d}x}{\mathrm{d}t} + \omega_0^2 x = f(t). \tag{2.1}$$

Next we take the Fourier transform of the equation of motion and perform partial integrations, which then yield

$$(-\mathrm{i}\omega)^2 X + \Gamma(-\mathrm{i}\omega) X + \omega_0^2 X = F, \tag{2.2}$$

where we have made the physically reasonable assumption that all the integrated parts vanish as $|t| \to \infty$. We can solve (2.2) for X and find a function already familiar in the context of the Lorentz model for the linear susceptibility,

$$X(\omega) = \frac{F(\omega)}{\omega_0^2 - \omega^2 - \mathrm{i}\Gamma\omega}. \tag{2.3}$$

The complex function of the real angular frequency variable in (2.3) is always finite, as we already know from previous treatment of this type of function. If complex angular frequencies are allowed we find simple poles in the lower half of the complex plane.

We now continue the mathematical description of the principle of causality by a procedure where an inverse Fourier transformation is applied to X and the integration order is changed, whereby the well-known convolution can be found as follows:

$$\begin{aligned} x(t) &= \frac{1}{2\pi} \int_{-\infty}^{\infty} X(\omega) \mathrm{e}^{-\mathrm{i}\omega t} \mathrm{d}\omega \\ &= \frac{1}{2\pi} \int_{-\infty}^{\infty} \frac{F(\omega) \mathrm{e}^{-\mathrm{i}\omega t}}{\omega_0^2 - \omega^2 - \mathrm{i}\Gamma\omega} \mathrm{d}\omega \\ &= \frac{1}{2\pi} \int_{-\infty}^{\infty} \mathrm{d}\omega \frac{\mathrm{e}^{-\mathrm{i}\omega t}}{\omega_0^2 - \omega^2 - \mathrm{i}\Gamma\omega} \int_{-\infty}^{\infty} f(t') \mathrm{e}^{\mathrm{i}\omega t'} \mathrm{d}t' \\ &= \int_{-\infty}^{\infty} f(t') \mathrm{d}t' \left(\frac{1}{2\pi} \int_{-\infty}^{\infty} \frac{\mathrm{e}^{-\mathrm{i}(t-t')}}{\omega_0^2 - \omega^2 - \mathrm{i}\Gamma\omega} \mathrm{d}\omega \right) \\ &= \int_{-\infty}^{\infty} f(t') g(t - t') \mathrm{d}t'. \end{aligned} \tag{2.4}$$

Here g is the Green's function, which is independent of the driving force, and g is expressed in the form

$$g(\tau) = \frac{1}{2\pi} \int_{-\infty}^{\infty} \frac{\mathrm{e}^{-\mathrm{i}\omega\tau}}{\omega_0^2 - \omega^2 - \mathrm{i}\Gamma\omega} \mathrm{d}\omega. \tag{2.5}$$

The Green's function is also called the response function of the system. We want to have an explicit expression for the Green's function. This is accomplished by the theorem of residues and Cauchy's integral theorem (Appendices A and B). In other words, we have to perform complex contour integrations in the complex angular frequency plane. Then we have to study the integral

$$I = \oint\limits_C \frac{\mathrm{e}^{-\mathrm{i}\widehat{\omega}\tau}}{\omega_0^2 - \widehat{\omega}^2 - \mathrm{i}\Gamma\widehat{\omega}} \mathrm{d}\widehat{\omega}, \tag{2.6}$$

where C is a closed contour.

The integration is possible using the two closed semicircular contours shown in Fig. 2.1. We have to split the calculation of $g(\tau)$ into two parts. One part is related to the case in which time is negative and the other one to the case in which time is positive. This choice of two time intervals has no effect on the fact that the poles of the integrand of (2.6) are always in the lower half plane. The choice of the integration contours as semicircles is a matter of convenience which makes the calculations of the related integrals easier. Other rectifiable closed Jordan curves, according to the homotopy theory of integration paths [25], will do equally well.

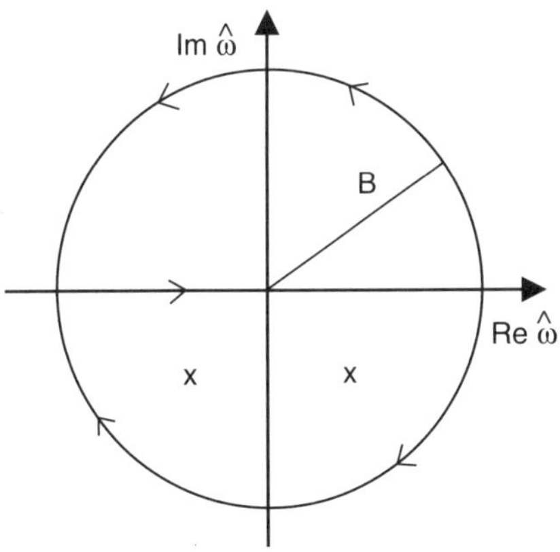

Fig. 2.1. Contours (semicircles above and below horizontal axis) for complex integrations for Green's function

When $\tau \geq 0$ we close the integration path in the lower half plane. In this case we can be sure that there will not be any exponential explosion of the complex function $\exp(-\mathrm{i}\widehat{\omega}\tau)$. Now the integrand is holomorphic almost everywhere, excluding the two poles. We can apply the theorem of residues, but we have to be sure that the integral along the arc of the semicircle vanishes as the radius B tends to infinity. This property, known as Jordan's lemma (Appendix C), can be shown to be valid by means of the following estimation,

$$\lim_{B\to\infty}\left|\int\limits_{|\hat{\omega}|=B,\mathrm{Im}\,\hat{\omega}<0}\frac{\mathrm{e}^{-\mathrm{i}\hat{\omega}\tau}}{\omega_0^2-\hat{\omega}^2-\mathrm{i}\Gamma\hat{\omega}}d\hat{\omega}\right|=0. \tag{2.7}$$

Indeed the integrand, G, has an upper bound

$$|G(\hat{\omega})|\leq\frac{1}{|\omega_0^2-\hat{\omega}^2-\mathrm{i}\Gamma\hat{\omega}|}\leq M(B),\quad \lim_{B\to\infty}M(B)=0. \tag{2.8}$$

The upper bound exists because the arc of the semicircle is a compact set and we are dealing with a sufficiently regular function. Now the theorem of residues states that

$$\begin{aligned}g(\tau)&=\frac{1}{2\pi}\oint\limits_{C}\frac{\mathrm{e}^{-\mathrm{i}\hat{\omega}\tau}}{\omega_0^2-\hat{\omega}^2-\mathrm{i}\Gamma\hat{\omega}}\mathrm{d}\hat{\omega}\\&=-\frac{1}{2\pi}2\pi\mathrm{i}\sum_{\text{poles}}\mathrm{Res}\left[\frac{\mathrm{e}^{-\mathrm{i}\hat{\omega}\tau}}{\omega_0^2-\hat{\omega}^2-\mathrm{i}\Gamma\hat{\omega}}\right]\\&=\mathrm{e}^{-(\Gamma/2)t}\frac{\sin\left\{\left[\omega_0^2-(\Gamma/2)^2\right]^{1/2}t\right\}}{\left[\omega_0^2-(\Gamma/2)^2\right]^{1/2}}.\end{aligned} \tag{2.9}$$

Similarly, when $\tau<0$, we close the integration contour in the upper half plane. Then the estimation of the integral along the arc of the semicircle is equivalent to that in (2.7). Furthermore, the integrand is a holomorphic function in the upper half plane. The application of Cauchy's theorem is possible and it means that

$$\oint\limits_{C}\frac{\mathrm{e}^{-\mathrm{i}\hat{\omega}\tau}}{\omega_0^2-\hat{\omega}^2-\mathrm{i}\Gamma\hat{\omega}}\mathrm{d}\hat{\omega}=0. \tag{2.10}$$

As a result of these two calculations we can write

$$g(\tau)=H(\tau)\,\mathrm{e}^{-(\Gamma/2)t}\frac{\sin\left\{\left[\omega_0^2-(\Gamma/2)^2\right]^{1/2}t\right\}}{\left[\omega_0^2-(\Gamma/2)^2\right]^{1/2}}, \tag{2.11}$$

where H is the Heaviside step function,

$$H(\tau)=\begin{cases}1, & \tau\geq 0\\ 0, & \tau<0.\end{cases} \tag{2.12}$$

We observe that the response function is real. However, in Raman emission profile spectroscopy [26, 27] and in nonlinear optical spectroscopy [28] complex response functions can be present. The origin of the complexity of the response functions is in the quantum mechanical treatment according to Dirac–Heisenberg models. After the above derivation we finally get the desired result,

$$x(t)=\int_{-\infty}^{t} f(t')\,\mathrm{e}^{-(\Gamma/2)(t-t')}\frac{\sin\left\{\left[\omega_0^2-(\Gamma/2)^2\right]^{1/2}(t-t')\right\}}{\left[\omega_0^2-(\Gamma/2)^2\right]^{1/2}}\mathrm{d}t'. \quad (2.13)$$

The interpretation of (2.13) is as follows. The upper limit of the integration is now t, which implies that the displacement $x(t)$ depends only on the driving force that acted in the past but not on its values in the future. Such a property is called the causality of the system. The choice of the damping parameter as a positive number defines the correct direction of time. Time would be a reversible quantity in the case of damping-free systems.

The principle of causality can be written for the time-dependent linear susceptibility,

$$\chi^{(1)}(\omega)=\int_{0}^{\infty}\chi^{(1)}(t)\,\mathrm{e}^{\mathrm{i}\omega t}\mathrm{d}t, \quad (2.14)$$

and the lower bound of the integration is zero because of the fact that the response of the system is always at a later instant than the disturbance that causes it. Note that we can replace the angular frequency variable by a complex variable $\hat{\omega}=\omega+\mathrm{i}v$ in the integral of (2.14). The convergence of the integral is improved, while at the same time the complex linear susceptibility as a function of the complex angular frequency turns out to be a holomorphic function in the upper half plane.

In the context of the general theory of causality, response functions and dispersion relations, square integrability [10] of the modulus of a complex function such as the linear susceptibility guarantees the existence of the dispersion relations, which are the topics of the forthcoming sections.

2.2 Hilbert Transforms

We have seen that the complex linear susceptibility, permittivity and complex refractive index are holomorphic functions in the upper half plane. Therefore some fundamental results of complex analysis hold for these quantities. At the beginning of the 20th century Kramers and Kronig derived dispersion relations which are of general validity for linear optical constants, and since then these relations have been widely used in theoretical and experimental studies in which optical data inversion is required.

Next we derive the so-called Hilbert transforms [29], which are the key factors that are needed in order to write down the Kramers–Kronig dispersion relations. In mathematical terms, the problem leading to the Hilbert transforms is closely related to the Dirichlet boundary value problem. The solution of the Dirichlet problem for the half plane is known as the Poisson formula [12, 13]. The Poisson formula makes it possible to construct a harmonic function in the upper half plane when its boundary value on the real axis is known.

Although we are dealing mostly with angular frequency variables, for the sake of generality, we use here a general complex variable, $z = x + \mathrm{i}y$, which can also denote complex angular frequency. In addition, the complex linear susceptibility, etc. are denoted by a complex function $f(x) = U(x) + \mathrm{i}V(x)$, which is assumed to be holomorphic in the upper half of the complex plane, $\mathrm{Im}z \geq 0$. The asymptotic fall-off required of f is as follows:

$$|f(z)| \leq \frac{\text{constant}}{B^{\delta}}, \quad |z| = B, \tag{2.15}$$

where $\delta > 0$.

This property is indeed fulfilled by the linear optical constants, as well as by the nonlinear susceptibilities that will be treated later in more detail. As we have seen, Newtonian mechanics and the Lorentz oscillator model yielded an asymptotic fall-off of the linear susceptibility proportional to ω^{-2} for high angular frequencies.

The derivation of the Hilbert transforms is based on introducing a singular point x' of a function g. Then, on the real axis, either $g(x') = \infty$ or $g(x') = -\infty$. However, an integral

$$I = \mathrm{P}\int_{-\infty}^{\infty} g(x)\,\mathrm{d}x, \tag{2.16}$$

where P denotes the Cauchy principal value, can exist. The Cauchy principal value is defined by a limiting process:

$$I = \lim_{\epsilon\to 0}\left\{\int_{-\infty}^{x'-\epsilon} g(x)\,\mathrm{d}x + \int_{x'+\epsilon}^{\infty} g(x)\,\mathrm{d}x\right\}, \tag{2.17}$$

where the singular point is approached simultaneously from left and right in a symmetric manner. Note carefully that taking the principal value, in the context of Hilbert transforms for optical constants, includes the approaching of plus and minus infinities also in a symmetric manner. This becomes evident after the following derivation, which uses a complex contour integration:

$$\mathrm{P}\int_{-\infty}^{\infty} \frac{f(x)}{x-x'}\mathrm{d}x = \lim_{\epsilon\to 0}\lim_{B\to\infty}\left\{\oint_C \frac{f(z)}{z-x'}\mathrm{d}z - \int_D \frac{f(z)}{z-x'}\mathrm{d}z - \int_A \frac{f(z)}{z-x'}\mathrm{d}z\right\}, \tag{2.18}$$

where the integration contours are as shown in Fig. 2.2.

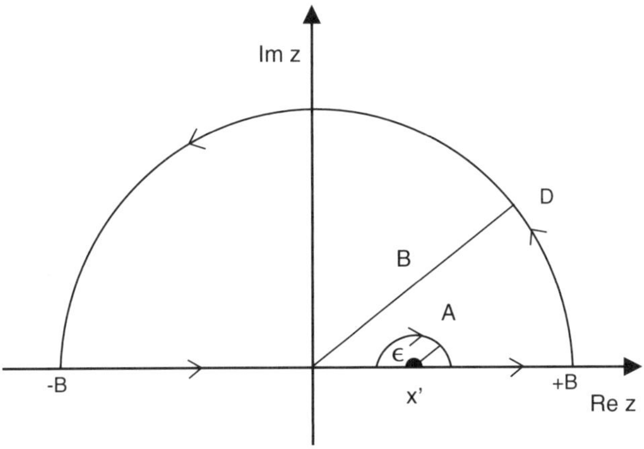

Fig. 2.2. Contours for derivation of Hilbert transforms

The singular point, which is a single pole on the real axis (along which, in reality, the numerical integrations are performed), is artificially avoided by making a small semicircular detour, whose radius ϵ is allowed to tend to zero. A closed contour is obtained by connecting the end points on the real axis, using the large semicircle whose radius B is allowed to approach infinity. We have to calculate the individual integrals on the right-hand side of (2.18). Since the function f is holomorphic in the upper half plane and there are no poles within the closed contour C, then according to Cauchy's integral theorem the first integral on the right-hand side of (2.18) is equal to zero. Next consider the integral along the arc of the large semicircle. The arc is a closed set and f, as a holomorphic function, is uniformly continuous on the arc. Therefore it will possess a maximum value

$$M(B) = \max_{|z|=B, \mathrm{Im}\, z \geq 0} |f(z)| \ . \tag{2.19}$$

Because of the assumption of the property expressed by (2.15), $\lim_{B\to\infty} M(B) = 0$. Now we can estimate the integral along the arc (an estimation based on Jordan's lemma) and find that

$$\left| \int_D \frac{f(z')}{z-x'} \mathrm{d}z \right| \leq \int_D \frac{|f(z)|\,|\mathrm{d}z|}{|z-x'|} = \int_D \frac{|f(z)|\,|\mathrm{d}z|}{|B\mathrm{e}^{\mathrm{i}\varphi} - x'|}$$

$$\leq M(B) \int_0^{\pi} \frac{B\mathrm{d}\varphi}{\sqrt{B^2 + x'^2 - 2Bx' \cos\varphi}}$$

$$\leq M(B) \int_0^\pi \frac{B}{B - x'} d\varphi$$

$$= \pi \frac{BM(B)}{B - x'} \to 0, \text{ as } B \to \infty, \tag{2.20}$$

where the estimate $\cos\varphi = 1$ has been used and x' is assumed to be positive. If x' is negative a similar estimate to that of (2.20) can be made, but now it is better to estimate that $\cos\varphi = -1$. The integration procedure in (2.20) involves a limiting process where the radius $B \to \infty$. Therefore the Cauchy principal-value integral includes not only the symmetric approaching of the singular point x', but also the symmetric approaching of plus and minus infinity. This fact has to be taken into account in numerical integrations performed by computer, when data inversion is done by means of Hilbert transforms. We remark that the existence of the Cauchy principal value does not imply the existence of the conventional nonessential integral, whereas the existence of a nonessential integral always implies its existence also as a Cauchy principal value.

There is one integration left, that is to say, the counterclockwise integration along the small semicircle. Now there is a single pole within the contour but the function is holomorphic within the integration path. We can apply the theorem of residues, and obtain the result

$$\mathrm{P} \int_{-\infty}^{\infty} \frac{f(x)}{x - x'} dx = i\pi f(x') . \tag{2.21}$$

We can now separate the real and imaginary parts:

$$U(x') = +\frac{1}{\pi} \mathrm{P} \int_{-\infty}^{\infty} \frac{V(x)}{x - x'} dx ,$$

$$V(x') = -\frac{1}{\pi} \mathrm{P} \int_{-\infty}^{\infty} \frac{U(x)}{x - x'} dx , \tag{2.22}$$

which are known as the Hilbert transforms. The advantage of these integral equations is that if U is known then V can be calculated and vice versa. The holomorphism of the optical constants was related to the causality. Therefore we can say that the Hilbert transforms for the optical constants are restatements of the principle of causality. Indeed the interaction of light with matter is completely known in linear optical spectroscopy when either the imaginary part or the real part of the optical constant is known.

As concerns the numerical integrations needed in (2.22), we cannot allow the integrals to have precisely the singular value. Reliable estimates are obtained for $U(x')$ and $V(x')$ when the values of the integrals cease to change for practical purposes while the symmetric approach to the point x' becomes closer and closer. The shortcoming of the Hilbert transforms for data inversion is that one integration will yield the desired physical parameter at only one point. This means that the point x' must be scanned over the infinite real axis in order to obtain the wanted quantity.

Another problem, which is always present with measured data, is concerned with data extrapolations beyond the measured range. The spectral data is limited and usually far away from the lower and upper bounds of the Hilbert transform integrals. Fortunately, in most cases, symmetry properties of the optical constants or susceptibilities allow us to replace the lower bound of integration by a physically more reasonable limit, zero. However, in the theory of Raman emission spectroscopy, the response function can be complex, which implies that Hilbert transforms [26] are not as useful in Raman emission spectroscopy as in other optical spectroscopies.

It is noteworthy that the Hilbert transform of a constant function yields a zero value; i.e. the Hilbert transforms of $U(x)$ and $U(x) + K$, where K is a constant, are the same. Therefore we obtain the same value for an integral where the integrand $U(x)/(x' - x)$ is replaced by $[U(x) - U(x')]/(x - x')$, for example. Obviously, the latter form approaches the derivative of U when x approaches x'. It is thus evident that by Hilbert transforms we may usually calculate a change in a physical parameter. The additive constant K that may be needed is usually known from the physics of the system, and it has a unique value. This is a necessary property in the derivation of the Hilbert transforms. That is to say, we must choose K, in accordance with the physics, to guarantee the vanishing of the complex-valued integral along the arc of the large semicircle when its radius tends to infinity. An example of the constant K is in the relation between the linear susceptibility and the permittivity (1.7), where $K = \varepsilon_0$. We observe from (1.7) that we cannot write Hilbert transforms for the permittivity since the condition of asymptotic fall-off of the permittivity required by (2.15) is not fulfilled. However, it is fulfilled for the function $\varepsilon - \varepsilon_0$, which is the linear susceptibility.

In optical physics, the variable x can be the time variable. Then the Hilbert transforms characterize an "analytic signal", the concept of which has had importance in the description of the coherence properties of optical fields, as described in the seminal paper [30] and in the book by Klauder and Sudarshan [31].

Other applications of Hilbert transforms are described in the book by Bohren and Huffman [32].

2.3 Kramers–Kronig Relations in Transmission Spectroscopy

Nowadays a standard spectroscopic measurement of transparent materials is the measurement of transmission, using a spectrophotometer, as a function of wavelength. The principle of the measurement is shown in Fig. 2.3.

Since the thickness of the sample can be measured rather easily, we may apply the Beer–Lambert law of (1.23) together with (1.24) in order to de-

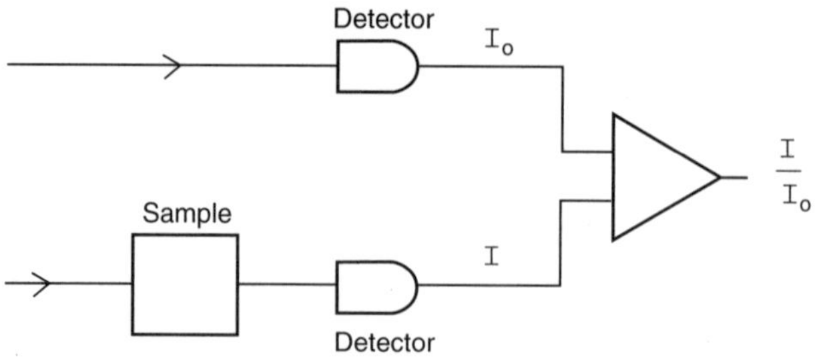

Fig. 2.3. Principle of transmission measurement

termine from the measured spectrum the imaginary part of the complex refractive index. Thereafter, we wish to obtain the change of the real refractive index. This can be accomplished using the appropriate Kramers–Kronig relation. The complex refractive index is a holomorphic function in the upper half plane, it has the required asymptotic properties for high angular frequencies, and therefore a Hilbert transform pair can be written that connects the real refractive index and the corresponding extinction coefficient. The Kramers–Kronig dispersion relations can be considered as modified Hilbert transforms. The Kramers–Kronig relations can be written after we recognize the symmetry properties

$$n(-\omega) = n(\omega),$$

$$k(-\omega) = -k(\omega), \tag{2.23}$$

which can be derived by using, for example, the definitions of n and k given by (1.26) and the symmetry properties stated in (1.14). As an example we derive the Kramers–Kronig relation for the real refractive index as follows:

$$\begin{aligned} n(\omega') - 1 &= \frac{1}{\pi} \mathrm{P} \int_{-\infty}^{\infty} \frac{k(\omega)}{\omega - \omega'} \mathrm{d}\omega \\ &= \frac{1}{\pi} \mathrm{P} \int_{0}^{\infty} \frac{k(\omega)}{\omega - \omega'} \mathrm{d}\omega - \frac{1}{\pi} \mathrm{P} \int_{0}^{-\infty} \frac{k(\omega)}{\omega - \omega'} \mathrm{d}\omega \\ &= \frac{1}{\pi} \mathrm{P} \int_{0}^{\infty} \frac{k(\omega)}{\omega - \omega'} \mathrm{d}\omega - \frac{1}{\pi} \mathrm{P} \int_{0}^{\infty} \frac{k(-\omega)}{-\omega - \omega'} (-\mathrm{d}\omega) \\ &= \frac{2}{\pi} \mathrm{P} \int_{0}^{\infty} \frac{\omega k(\omega)}{\omega^2 - \omega'^2} \mathrm{d}\omega. \end{aligned} \tag{2.24}$$

After a similar derivation for the extinction coefficient, we get the two famous Kramers–Kronig relations as follows:

$$n(\omega') - 1 = \frac{2}{\pi} \mathrm{P} \int_0^\infty \frac{\omega k(\omega)}{\omega^2 - \omega'^2} \mathrm{d}\omega,$$

$$k(\omega') = -\frac{2\omega'}{\pi} \mathrm{P} \int_0^\infty \frac{n(\omega') - 1}{\omega^2 - \omega'^2} \mathrm{d}\omega, \tag{2.25}$$

which hold for linearly polarized light modes in isotropic materials. Similar relations, but of more general validity and which take into account the tensor nature of the permittivity of anisotropic materials [33], can be written for the complex permittivity and linear susceptibility of insulators, metals and semiconductors. However, the measurement of the permittivity is problematic; therefore it is more convenient to apply the Kramers–Kronig relation, whenever possible, for corresponding optical constants that are easier to measure.

In the practical use of the Kramers–Kronig relations related to physical or technological studies of either "well-known" or novel materials, care should be taken when extrapolating the measured data at the "wings" of the measured spectra into regions where the optical constant is unknown. Especially, the data extrapolation must be consistent with the symmetry properties of (2.23). This is illustrated by the following example, where a Gaussian-shaped function is chosen to describe the extinction coefficient [34], i.e.

$$k(\omega) = A\mathrm{e}^{-[(\omega - \omega_0)/W]^2}, \tag{2.26}$$

where A is the amplitude, ω_0 is the central angular frequency and $2(\ln 2)^{1/2}W$ is the full width at half maximum. We observe immediately that $k(-\omega) \neq -k(\omega)$. Now, the calculation of the refractive index change by means of the Kramers–Kronig relation can be hazardous, since the assumption of the symmetry property imposed on the extinction coefficient is not valid. The correct refractive index change is obtained from the corresponding Hilbert transform for the real refractive index change, where the integration range is now $(-\infty, \infty)$ instead of the range $(0, \infty)$. It can be observed in this case that when using an incorrect Kramers–Kronig representation, the ratio ω_0/W is crucial in the estimation of the correct values of the refractive index change. If ω_0/W is large, then the Kramers–Kronig relation gives a good approximation. On the other hand, if the ratio has a relatively low value, then (2.25) gives erroneous result, as demonstrated in Fig. 2.4. It is worth noting that the frequency at which the dispersion is zero is shifted as ω_0/W decreases.

Kramers–Kronig analysis can be exploited to investigate dynamic changes of refractive index. As an example of such a procedure consider the system illustrated in Fig. 2.5. It describes ruby-laser-light-induced temporary memory, based on the use of a lithium-doped KCl crystal that contains F_A color centers (for more details about F_A and other color centers see e.g. [35]). Linearly polarized light pulses (pulse duration 25 ns and time lag between successive

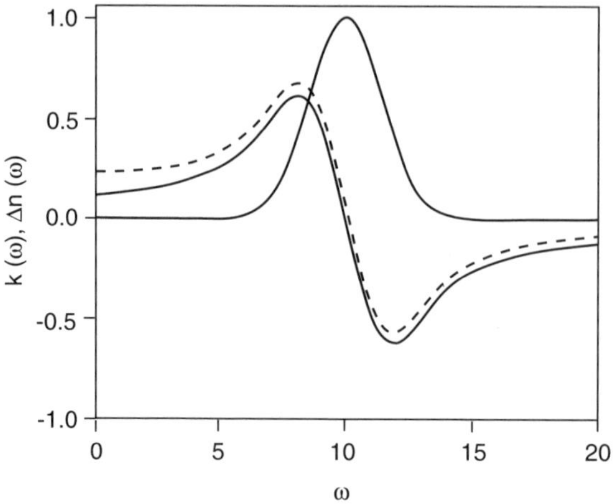

Fig. 2.4. Example of extinction curve and the corresponding true (*solid line*) and erroneous (*dashed line*) dispersion curves

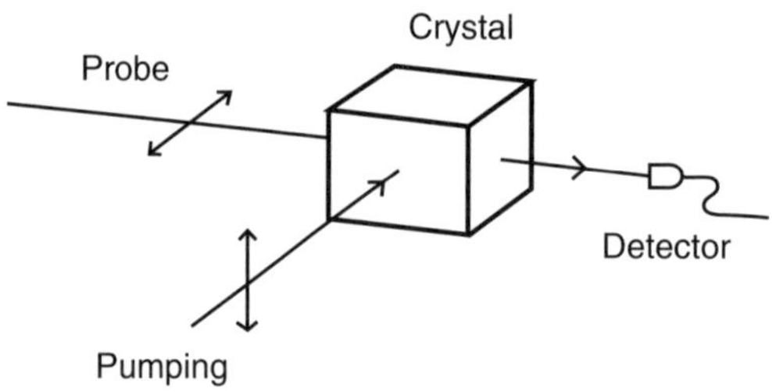

Fig. 2.5. Schematic diagram of pumping and probing of KCl:Li$^+$ crystal

pulses 25 s) from a ruby laser cause reorientation of the F_A centers and simultaneously there appears a conversion of F_A centers to aggregate color centers. As a result of continuous pumping by the ruby laser, and the induced color center conversion, the extinction coefficient, as well as the refractive index of the crystal, changes gradually as a function of pumping time. The optical density of the crystal was measured using linearly polarized, weak-intensity probe light from a spectrophotometer. The change of the refractive index was calculated, with proper data extrapolations, using the Kramers–Kronig relation [36]. The results of the calculations are shown in Fig. 2.6.

From (2.25) we can derive a specific dispersion relation that gives us information about the static or DC refractive index change:

$$n(0) - 1 = \frac{2}{\pi} \mathrm{P} \int_0^\infty \frac{k(\omega)}{\omega} \mathrm{d}\omega. \tag{2.27}$$

The integral can exist also as a conventional nonessential integral, depending on the behavior of $k(\omega)$ as $\omega \to 0$. The relation (2.27) is valid only for

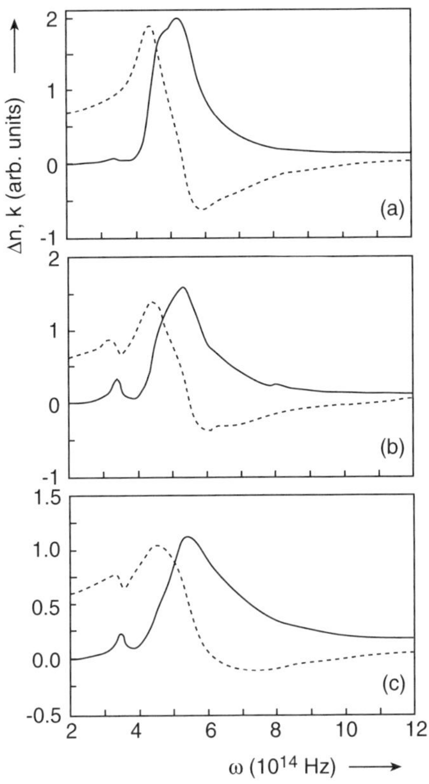

Fig. 2.6. Extinction coefficient (*solid line*) and refractive index change (*dashed line*) of KCl:Li$^+$ crystal as a function of frequency **(a)** 0 h, **(b)** 38 h and **(c)** 92 h pumping

insulators. In the case of metals both n and k behave as $\omega^{-1/2}$ when $\omega \to 0$. However, if we write relation (2.27) for the linear susceptibility, it holds for both insulators and metals. This is because with metals $\mathrm{Im}\{\chi^{(1)}\}$ behaves like ω^{-1} but $\mathrm{Re}\{\chi^{(1)}\}$ remains finite. The static or low-frequency value of the permittivity of an insulator has importance, for example, in electric power applications, where the properties of the insulating materials of capacitors are important.

2.4 Multiply-Subtractive Kramers–Kronig Relations

In the theory of causal response functions it is required that the response function in Fourier space is square integrable. This may not, however, be always the case, but square integrability can be achieved using the appropriate subtraction technique, as given by Nussenzveig [10].

Another aspect of the subtraction technique, not related to the lack of square integrability of optical constants, is the possibility to get better estimates in data inversion. The purpose of the subtractive technique in data inversion is to create a mathematical model in which the data extrapolations beyond the measured spectral range are not as crucial as in the application of the conventional Kramers–Kronig relations. This means a compromise with the number of measurements. In other words, if we are to calculate the change of refractive index, then we have to know the refractive index at some discrete set of reference or "anchor" angular frequencies. The reference points should be preferably in the spectral range of the measured absorption. Ahrenkiel [37] presented a singly-subtractive Kramers–Kronig relation, which has the form

$$\begin{aligned} n(\omega') - n(\omega'') &= [n(\omega') - 1] - [n(\omega'') - 1] \\ &= \frac{2}{\pi} \mathrm{P} \int_0^\infty \frac{\omega k(\omega)}{\omega^2 - \omega'^2} \mathrm{d}\omega - \frac{2}{\pi} \mathrm{P} \int_0^\infty \frac{\omega k(\omega)}{\omega^2 - \omega''^2} \mathrm{d}\omega \\ &= \frac{2}{\pi} \left(\omega'^2 - \omega''^2\right) \mathrm{P} \int_0^\infty \frac{\omega k(\omega)}{\left(\omega^2 - \omega'^2\right)\left(\omega^2 - \omega''^2\right)} \mathrm{d}\omega, \end{aligned} \tag{2.28}$$

for the refractive index change. The advantage of this type of integral is that it converges more rapidly than the conventional Kramers–Kronig integral. Now the value of the refractive index at an anchor point, ω'', has to be measured somehow, while ω' is the variable.

In order to improve the accuracy of data inversion a multiply-subtractive technique has been introduced by Goplen et al. [38]. In their procedure they limit the integration interval to the interval of the measured data, $(\omega_\mathrm{i}, \omega_\mathrm{f})$. In addition it is defined that

$$\overline{n} = n(\omega') - \frac{2}{\pi} \mathrm{P} \int_{\omega_\mathrm{i}}^{\omega_\mathrm{f}} \frac{\omega k(\omega)}{\omega^2 - \omega'^2} \mathrm{d}\omega, \tag{2.29}$$

where $\overline{n}$ replaces the high-frequency value of the refractive index in order to take into account the loss of contributions from the wings of the extinction curve. Next it is assumed that the refractive index is known over a discrete set containing L reference points, of angular frequencies $(\omega_1, ..., \omega_L)$. In the case of liquids this additional information about the refractive index can be obtained using attenuated total reflection (ATR) measurements, whereas for solid materials ellipsometric measurements using laser lines are possible. Under the above assumptions, we can write

$$\overline{n} = n(\omega_1) - \frac{2}{\pi} \mathrm{P} \int_{\omega \mathrm{i}}^{\omega_\mathrm{f}} \frac{\omega k(\omega)}{\omega^2 - \omega_1^2} \mathrm{d}\omega,$$

$$\overline{n} = n(\omega_2) - \frac{2}{\pi} \mathrm{P} \int_{\omega \mathrm{i}}^{\omega_\mathrm{f}} \frac{\omega k(\omega)}{\omega^2 - \omega_2^2} \mathrm{d}\omega,$$

$$\vdots \qquad \vdots$$

$$\overline{n} = n(\omega_L) - \frac{2}{\pi} \mathrm{P} \int_{\omega \mathrm{i}}^{\omega_\mathrm{f}} \frac{\omega k(\omega)}{\omega^2 - \omega_L^2} \mathrm{d}\omega \tag{2.30}$$

and solve for $\overline{n}$:

$$\overline{n} = \frac{1}{L} \left[\sum_{j=1}^{L} \left(n(\omega_j) - \frac{2}{\pi} \mathrm{P} \int_{\omega \mathrm{i}}^{\omega_\mathrm{f}} \frac{\omega k(\omega)}{\omega^2 - \omega_j^2} \mathrm{d}\omega \right) \right]. \tag{2.31}$$

Finally, when we substitute the result of (2.31) into (2.29), we find the multiply-subtractive Kramers–Kronig relation

$$n(\omega') = \frac{1}{L} \left[\sum_{j=1}^{L} \left(n(\omega_j) - \frac{2}{\pi} \mathrm{P} \int_{\omega_\mathrm{i}}^{\omega_\mathrm{f}} \frac{\omega k(\omega)}{\omega^2 - \omega_j^2} \mathrm{d}\omega \right) \right]$$

$$+ \frac{2}{\pi} \mathrm{P} \int_{\omega_\mathrm{i}}^{\omega_\mathrm{f}} \frac{\omega k(\omega)}{\omega^2 - \omega'^2} \mathrm{d}\omega. \tag{2.32}$$

Another type of method to analyse optical data in a finite frequency interval is the one given by Hulthen [39]. Without going into the details of this method, we briefly mention that in principle it allows the calculation of the real refractive index and the extinction coefficient for all frequencies provided that the optical constants are known for at least partly overlapping angular frequency intervals.

2.5 Imaginary Angular Frequencies

In physical reality, we always are restricted by the fact that when measuring optical spectra, the angular frequency is positive. Negative and purely imaginary values of the angular frequency are beyond the scope of any experiment. However, we have a desire to find rather general theoretical methods for spectral analysis. For this purpose it is possible to introduce angular frequency values that may yield new information about spectra. In this section we investigate what happens if we choose a purely imaginary angular frequency as a variable for the optical constants.

Let us first study the linear susceptibility of the Lorentzian model (1.10) in the complex angular frequency plane, where $\hat{\omega} = \mathrm{i}\omega$. Then we find that

$$\chi^{(1)}(\omega) = \frac{\aleph e^2}{m\varepsilon_0} \frac{1}{\omega_0^2 + \omega^2 + \Gamma\omega}, \tag{2.33}$$

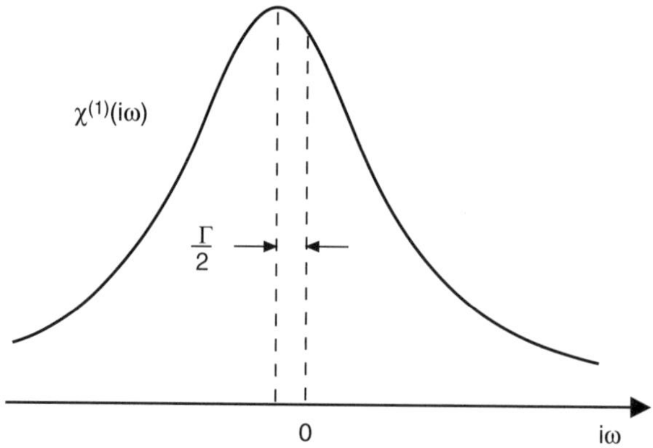

Fig. 2.7. Curve illustrating the function $\chi^{(1)}\,(\mathrm{i}\omega)$

where $\chi^{(1)}\,(\mathrm{i}\omega)$ has now to be a real function, as illustrated in Fig. 2.7, of the imaginary variable. From (2.33) we observe immediately that the linear susceptibility is neither an even nor an odd function of the imaginary variable. In the the case of the Lorentzian model we can make the approximation that $\chi^{(1)}(0\ \mathrm{s}^{-1}) \simeq \chi^{(1)}(\mathrm{i}\ \mathrm{s}^{-1})$. Note that the poles of this new function are now located in the left half of the complex angular frequency plane. Furthermore, it is possible that the poles are located on the negative real axis. Nevertheless, the function of (2.33) is a holomorphic function of the variable $\mathrm{i}\hat{\omega}$ in the right-hand half plane.

If we do not restrict the linear susceptibility to that of the Lorentzian model, but apply the more general response function of (2.14), we find that

$$\chi^{(1)}\,(\mathrm{i}\omega) = \int_0^\infty \chi^{(1)}\,(t)\,\mathrm{e}^{-\omega t}\mathrm{d}t. \qquad (2.34)$$

This is nothing but an integral of Laplace-transform type, which would diverge if we allowed negative values of the time variable, and also if we allowed positive time values but negative ω (i.e. if we had chosen $-\mathrm{i}\omega$ instead of $+\mathrm{i}\omega$).

It is possible to find other types of dispersion relations than the Kramers–Kronig ones for a function that is holomorphic in the upper half plane and possesses an asymptotic fall-off for high angular frequencies in the complex plane as is the case for the complex linear susceptibility and optical constants. For the sake of generality we exploit again the "anonymous" holomorphic complex function $f(z) = U(x,y) + \mathrm{i}V(x,y)$, where $f(x,0) = U(x,0) + \mathrm{i}V(x,0)$ describes, for example, the optical constants, which are restricted to the real variable x (angular frequency variable). According to complex analysis, we can construct a complex function in the upper half plane $\mathrm{Im}\, z \geq 0$, provided that we know the function on the real axis (the boundary of the upper half plane). This can be put into the language of mathematics as

$$f(\varsigma) = \frac{1}{\pi} \int_{-\infty}^{\infty} \frac{\eta f(x)}{(x-\xi)^2 + \eta^2} \mathrm{d}x, \tag{2.35}$$

where $\varsigma = \xi + \mathrm{i}\eta$ is any point in the upper half plane; this is known as the Poisson formula. If we separate the real and imaginary parts, we find a pair of equations, known as the Poisson's formulae for the half plane, as follows:

$$U(\xi,\eta) = \frac{1}{\pi} \int_{-\infty}^{\infty} \frac{\eta U(x,0)}{(x-\xi)^2 + \eta^2} \mathrm{d}x,$$

$$V(\xi,\eta) = \frac{1}{\pi} \int_{-\infty}^{\infty} \frac{\eta V(x,0)}{(x-\xi)^2 + \eta^2} \mathrm{d}x. \tag{2.36}$$

Next, we apply this model to the case of complex linear susceptibility by choosing the point in the upper half plane so that $\varsigma = 0 + \mathrm{i}\eta$. The integrands in (2.36) have appropriate symmetry properties, and the second integral gives zero, in accordance with the requirement imposed on the linear susceptibility, $\mathrm{Im}\left\{\chi^{(1)}(\mathrm{i}\omega)\right\} = 0$. The first integral yields

$$\begin{aligned} \mathrm{Re}\left\{\chi^{(1)}(0,\eta)\right\} &= \frac{1}{\pi} \int_{-\infty}^{\infty} \frac{\eta \,\mathrm{Re}\left\{\chi^{(1)}(x,0)\right\}}{x^2+\eta^2} \mathrm{d}x \\ &= \frac{2}{\pi} \int_{0}^{\infty} \frac{\eta \,\mathrm{Re}\left\{\chi^{(1)}(x,0)\right\}}{x^2+\eta^2} \mathrm{d}x. \end{aligned} \tag{2.37}$$

The relation of (2.37) achieves a more practical form when we replace the variable x by the familiar angular frequency ω. The result (2.37) can be used, for example, to test theoretical models developed for the real susceptibility or real refractive index (for the extinction coefficient, $k(\mathrm{i}\omega) = 0$). The real refractive index change, where we "forget" the possibility of the real angular frequency variable ξ, is

$$n(\mathrm{i}\omega') - 1 = \frac{2\omega'}{\pi} \int_{0}^{\infty} \frac{n(\omega) - 1}{\omega^2 + \omega'^2} \mathrm{d}\omega. \tag{2.38}$$

Thus we could reconstruct a function that has no particular symmetry properties, using an integrand that certainly has fixed symmetry properties.

Another relation can be obtained using the complex contour integration procedure introduced by King [40]. Consider a function $[N(\hat{\omega}) - 1] \,/\, (\hat{\omega} + \mathrm{i}\omega')$, which is a holomorphic function in the upper half plane when ω' is real and greater than zero. Now, integrating along a closed semicircular contour in the upper half plane and making use of Cauchy's integral theorem and Jordan's lemma, we find that

$$\int_{-\infty}^{\infty} \frac{N(\omega) - 1}{\omega + \mathrm{i}\omega'} \mathrm{d}\omega = 0. \tag{2.39}$$

The symmetry properites of the complex refractive index make it possible to solve the following integral equation:

$$\int_{0}^{\infty} \frac{\omega k(\omega)}{\omega^2 + \omega'^2} \mathrm{d}\omega = \omega' \int_{0}^{\infty} \frac{n(\omega) - 1}{\omega^2 + \omega'^2} \mathrm{d}\omega. \tag{2.40}$$

Comparison of (2.40) with (2.38) provides us with new information, namely

$$n(\mathrm{i}\omega') - 1 = \frac{2}{\pi} \int_{0}^{\infty} \frac{\omega k(\omega)}{\omega^2 + \omega'^2} \mathrm{d}\omega. \tag{2.41}$$

We can continue the development by integrating both sides of (2.41) with respect to ω', and obtain

$$\int_{0}^{\infty} [n(\mathrm{i}\omega') - 1] \mathrm{d}\omega' = \frac{2}{\pi} \int_{0}^{\infty} \mathrm{d}\omega' \int_{0}^{\infty} \frac{\omega k(\omega)}{\omega^2 + \omega'^2} \mathrm{d}\omega = \int_{0}^{\infty} k(\omega) \mathrm{d}\omega, \tag{2.42}$$

a result that holds also for the linear permittivity.

2.6 Kramers–Kronig Relations in Reflection Spectroscopy

The purpose of a reflectometer is to detect specular reflection of light from bulk materials or thin films. The setup for reflectance measurement of solid materials is shown schematically in Fig. 2.8., and the setup for liquids [41] is shown in Fig. 2.9. Both techniques can provide spectra that can be inverted using dispersion relations.

During the 20th century the investigation of the wavelength dependence of reflection from opaque materials, such as metals and semiconductors, has provided a means to gain information about their optical constants. An important step in the estimation of the optical constants from reflectance spectra was the realization of the applicability of the Kramers–Kronig relations to the case of complex reflectance. The rigorous validity of Kramers–Kronig relations for normal incidence complex reflectance was studied by Velický [42], who examined the holomorphicity of the complex reflectance in the upper half of the complex angular frequency plane.

In the case of normal reflectance the complex electric-field reflectivity can be given in polar form with the aid of the complex refractive index:

$$r(\omega) = |r(\omega)| \mathrm{e}^{\mathrm{i}\varphi(\omega)} = \frac{N(\omega) - 1}{N(\omega) + 1}, \tag{2.43}$$

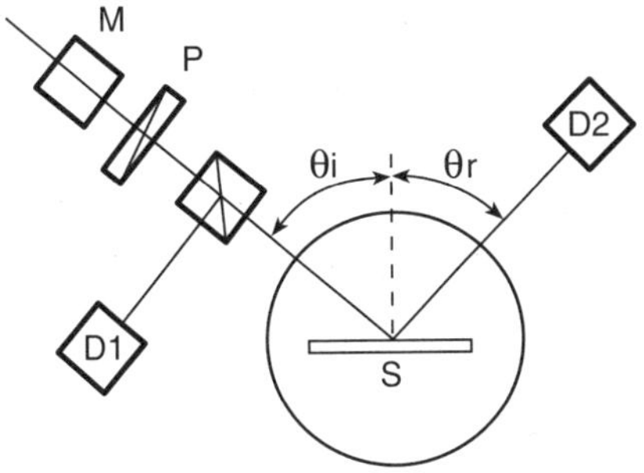

Fig. 2.8. Reflectometer used for reflectance measurement of solid materials. M = mirror, P = polarizer, D = detector and S = sample

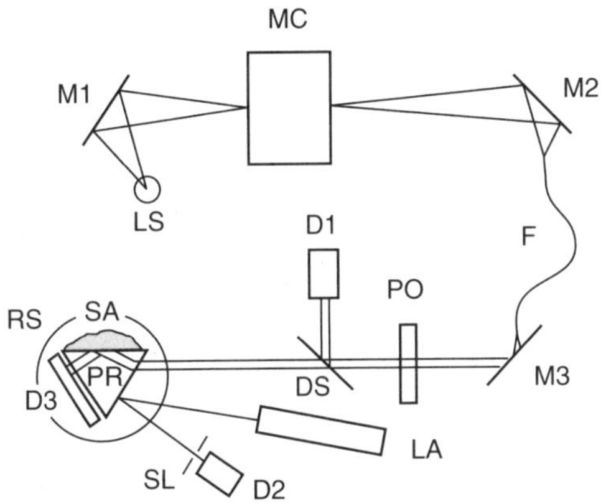

Fig. 2.9. Measurement of reflectance of liquids. M = mirror, MC = monochromator, D = detector, LA = laser, LS = light source, PO = polarizer, F = fiber, PR = prism and SL = slit

and r is a holomorphic function in the upper half plane. The symmetry property of the complex refractive index for complex angular frequency implies a symmetry property for the complex reflectivity as follows:

$$r\left(\hat{\omega}\right)^{*} = r\left(-\hat{\omega}^{*}\right). \tag{2.44}$$

Now taking the natural logarithm on both sides of (2.43) makes it possible to resolve the phase of the reflectivity and the logarithm of its modulus:

$$\ln[r\left(\omega\right)] = \ln\left|r\left(\omega\right)\right| + \mathrm{i}\varphi\left(\omega\right). \tag{2.45}$$

Using the symmetry property above and the definition (2.45), we find that the symmetry properties of the modulus and phase of the complex reflectivity are

$$\ln\left|r\left(-\omega\right)\right| = \ln\left|r\left(\omega\right)\right|,$$

$$\varphi(-\omega) = -\varphi(\omega). \tag{2.46}$$

The use of log $r(\omega)$ as ω tends to infinity, which in turn means that $r(\omega)$ tends to zero, seems to be problematic because of the logarithmic divergence of our function. The function of (2.45) is not square integrable along the real axis. Therefore we may expect to face problems that were not present when dealing with the complex refractive index. However, Kramers–Kronig dispersion relations can also be written for the logarithm of the complex reflectivity and it will be our next task to show this. If we can calculate the phase using the modulus data, then it is obvious that (2.43) will provide us with information about the optical constants. This kind of procedure where the phase is obtained from intensity data is called a phase retrieval procedure. Although Velický's proof is valid, we prefer here the proof given by Smith [43], which is based on the inspection of two different complex functions. The first function is

$$f(\hat{\omega}) = \frac{\ln[r(\hat{\omega})]}{\hat{\omega}^2 - \omega'^2}, \tag{2.47}$$

which is a holomorphic function almost everywhere in the upper half plane except at the two poles, $-\omega'$ and $+\omega'$, which are located on the real axis. In the case of conductors, there is a branch point at zero angular frequency, but this leads to no complication in the following derivation. Now we apply a complex contour integration similar to that used in the derivation of the Hilbert transforms. In other words we take a closed contour consisting of a semicircle at infinity in the upper half plane and (for insulators) the real axis from minus infinity to plus infinity, with infinitesimal semicircles to avoid the two poles (in the case of a metal the branch point is also avoided by an infinitesimal contour), as shown in Fig. 2.10.

Once again we apply the Cauchy principal value, Cauchy's integral theorem, Jordan's lemma $\left(f(\omega) \propto \omega^{-2}, \text{ as } \omega \to \infty\right)$ and the theorem of residues. Then we find that

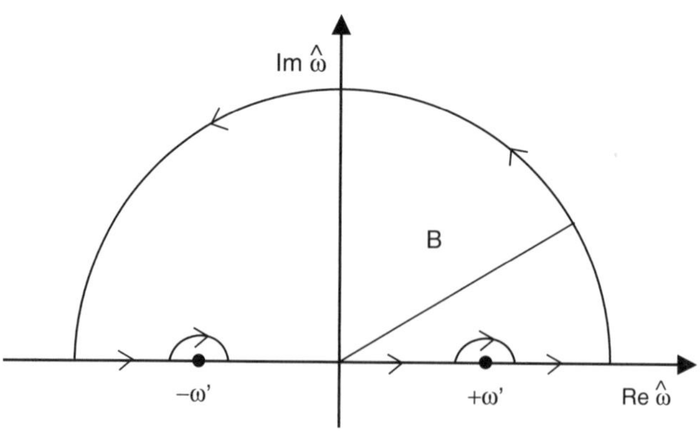

Fig. 2.10. Integration contour excluding singular points

$$\mathrm{P}\int_{-\infty}^{\infty} \frac{\ln[r(\omega)]}{\omega^2-\omega'^2}\mathrm{d}\omega$$

$$= \mathrm{i}\pi\left(\frac{1}{2\omega'}\left\{\ln|r(\omega')| - \ln|r(-\omega')| + \mathrm{i}\left[\varphi(\omega') - \varphi(-\omega')\right]\right\}\right). \quad (2.48)$$

Finally we make use of the symmetry properties of (2.46) and equate the real and imaginary parts of (2.48). Then we can resolve two integrals, one of them constituting one of the Kramers–Kronig relations for the complex reflectivity, as follows:

$$\mathrm{P}\int_{-\infty}^{\infty} \frac{\varphi(\omega)}{\omega^2-\omega'^2}\mathrm{d}\omega = 0,$$

$$\varphi(\omega') = -\frac{2\omega'}{\pi}\mathrm{P}\int_{-\infty}^{\infty} \frac{\ln|r(\omega)|}{\omega^2-\omega'^2}\mathrm{d}\omega. \quad (2.49)$$

The first of the above equations is a restatement of the symmetry property of the phase angle, whereas the second equation is a Kramers–Kronig relation, which allows the calculation of the phase angle. Strictly speaking, in the general case, there is an indeterminacy in the phase angle (recall what we stated about this subject in the context of Hilbert transforms) of an additive constant phase in the Kramers–Kronig relation of (2.49). However, the constant phase can be set to zero for both insulators and metals, since for such media at zero angular frequency, $\varphi\left(0\ \mathrm{s}^{-1}\right) = 0$.

Let us proceed to find the other part of the Kramers–Kronig relations. It can be obtained by inspection of the second complex function

$$f(\hat{\omega}) = \hat{\omega}\ln[r(\hat{\omega})]\left(\frac{1}{\hat{\omega}^2-\omega'^2} - \frac{1}{\hat{\omega}^2-\omega''^2}\right), \quad (2.50)$$

which is holomorphic almost everywhere in the upper half plane. We apply complex contour integration (now with four poles on the real axis) and, after a similar mathematical argument to that above, we find that

$$\mathrm{P}\int_{-\infty}^{\infty} \omega\ln[r(\omega)]\left(\frac{1}{\omega^2-\omega'^2} - \frac{1}{\omega^2-\omega''^2}\right)\mathrm{d}\omega$$

$$= \mathrm{i}\pi\sum_{\text{poles}}\mathrm{Res}\left[\hat{\omega}\ln[r(\hat{\omega})]\left(\frac{1}{\hat{\omega}^2-\omega'^2} - \frac{1}{\hat{\omega}^2-\omega''^2}\right)\right]. \quad (2.51)$$

After resolving the imaginary parts from both sides of (2.51), we find the missing Kramers–Kronig relation, which can be expressed as

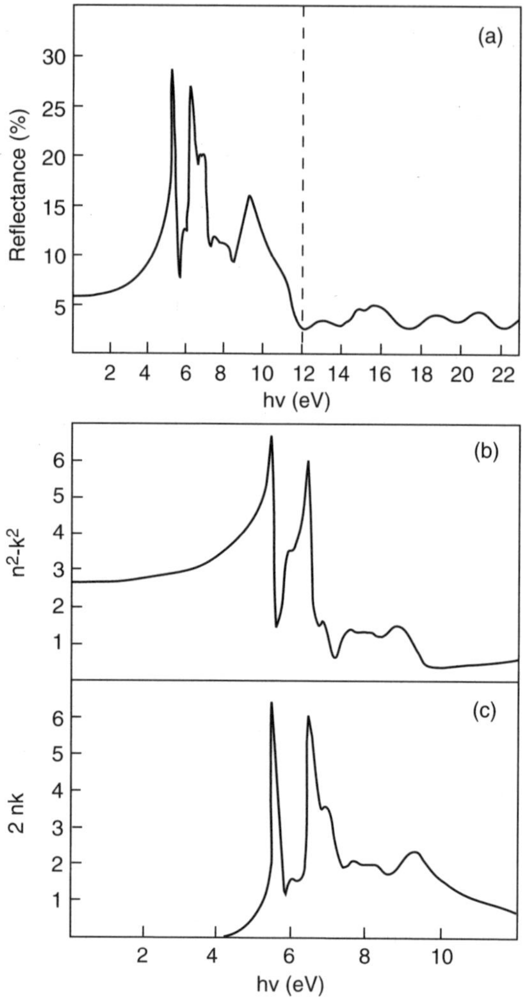

Fig. 2.11. **(a)** Reflectance spectrum of potassium iodide and **(b)** real and **(c)** imaginary parts of permittivity resolved by Kramers–Kronig analysis [44]

$$\ln |r(\omega')| - \ln |r(\omega'')| = \frac{2}{\pi} \mathrm{P} \int_0^\infty \omega\varphi(\omega) \left(\frac{1}{\omega^2 - \omega'^2} - \frac{1}{\omega^2 - \omega''^2} \right) \mathrm{d}\omega. \qquad (2.52)$$

There are a couple of things that should be carefully noted in the context of the Kramers–Kronig dispersion relation given in (2.52). First of all, it holds for both insulators and metals, but in the case of metals $\omega' \neq 0$ and $\omega'' \neq 0$. The dispersion relation in (2.52) is of subtractive form, where a reference point is used. Therefore, the reflectance at a reference or anchor point has to be known somehow, for instance from a single reflectance measurement with an appropriate laser line. Note that in the dispersion relation of (2.52) we cannot split the integral into two different Cauchy principal values, since what we then would obtain would be two divergent integrals. Convergence is only obtained using the subtractive form of (2.52).

In optical-reflectance data inversion we face again the problem of data extrapolations beyond the measured spectra. Therefore care has to be taken to fit the wings of the reflectance spectra so that the fitting is in accordance with the symmetry properties.

As an example of a Kramers–Kronig phase analysis we show in Fig. 2.11 the reflectance of potassium iodide and the corresponding real and imaginary parts of the permittivity which were calculated using the phase retrieval procedure.

The convergence of the integrals in the phase retrieval procedure can be improved using the technique of multiple-subtractive dispersion relations, just as in the case of the complex refractive index.

In the case of oblique incidence the Kramers–Kronig phase retrieval procedure can be exploited as well. However, some algebra is needed to resolve the optical constants. The formulae needed in oblique-incidence reflection spectroscopy for s-polarized light are derived in Appendix D.

In the general theory of phase retrieval [10, 24] it is possible that the logarithm of the modulus of a complex function has zeros in the upper half plane. Then the corresponding dispersion relations have to be revised in order to include terms that are due to a Blaschke product. Lee [27] has investigated such a possibility in the analysis of Raman emission profile spectra, where the so-called non-minimum phase retrieval problem may rise. Toll [24] has stated that zeros in the upper half plane introduce phase jumps of $2\pi M$, where M is an integer.

2.7 Kramers–Kronig Relations for the Effective Optical Constants of Two-Phase Nanocomposites

The optical properties of nanocomposites, and especially their nonlinear optical properties [45–50] have drawn much interest quite recently. Here, we concentrate on two standard models that have been applied in the description of nanocomposites. These are the layered-composite and Maxwell–Garnett models. Before going into details we write down the Wiener inequalities of the complex effective permittivity ε_{eff} of a multiphase system:

$$\left| \frac{1}{\sum\limits_{j=1}^{J} \left[f_j / \varepsilon_j(\hat{\omega}) \right]} \right| \leq \left| \varepsilon_{\text{eff}}(\hat{\omega}) \right| \leq \left| \sum_{j=1}^{J} f_j \varepsilon_j(\hat{\omega}) \right| , \tag{2.53}$$

where ε_j is the permittivity of the jth component, f_j is the volume content of the component and J is the number of components in the system. Equation (2.53) may help us to estimate limits for the effective permittivity of a multiphase system. The effective complex refractive index, for nonmagnetic materials, is defined now by

$$N_{\text{eff}}^2(\hat{\omega}) = \varepsilon_{\text{eff}}(\widehat{\omega}). \tag{2.54}$$

Let us first consider a layered two-phase nanocomposite by making the assumption that the scattering of light is negligible. In other words we assume that the features of the system are small compared to the wavelength. Such a sandwich system is illustrated in Fig. 2.12. We consider p-polarized light and, in particular, the component perpendicular to the plane of the layers.

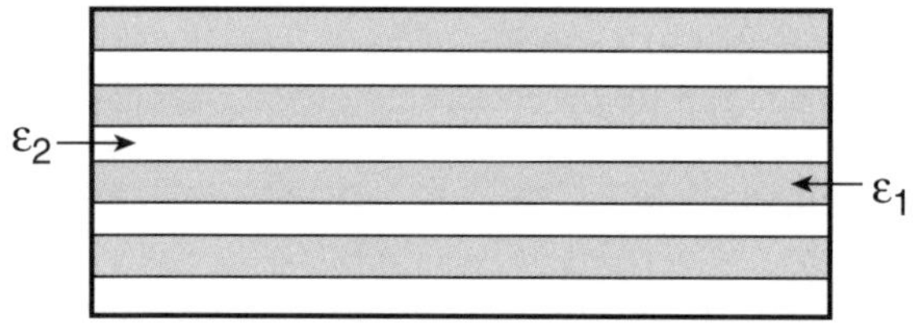

Fig. 2.12. Model of layered two-phase nanocomposite

The effective complex permittivity is obtained, for real angular frequency, as follows:

$$\frac{1}{\varepsilon_{\text{eff}}(\omega)} = \frac{f_1}{\varepsilon_1(\omega)} + \frac{f_2}{\varepsilon_2(\omega)}, \qquad f_1 + f_2 = 1. \tag{2.55}$$

The effective linear susceptibility

$$\chi_{\text{eff}}^{(1)}(\omega) = \varepsilon_{\text{eff}}(\omega) - 1$$

$$= \frac{\varepsilon_1(\omega)[\varepsilon_2(\omega) - 1] + f_1[\varepsilon_1(\omega) - \varepsilon_2(\omega)]}{\varepsilon_1(\omega) + f_1[\varepsilon_2(\omega) - \varepsilon_1(\omega)]} \tag{2.56}$$

is a holomorphic function in the upper half plane (it can posses zeros) and it has an appropriate asymptotic fall-off at high angular frequency. Therefore Kramers–Kronig relations for the effective complex refractive index can be obtained.

Another example of a two-phase nanocomposite material is a system described by the Maxwell–Garnett model. Maxwell–Garnett [51, 52] considered, at the beginning of the 20th century, the effective optical properties of a medium containing minute metal spheres. He was interested in the colors of glasses containing metal spheres. In his model, spherical inclusions are distributed in a host material in a uniform manner. The system is illustrated in Fig. 2.13.

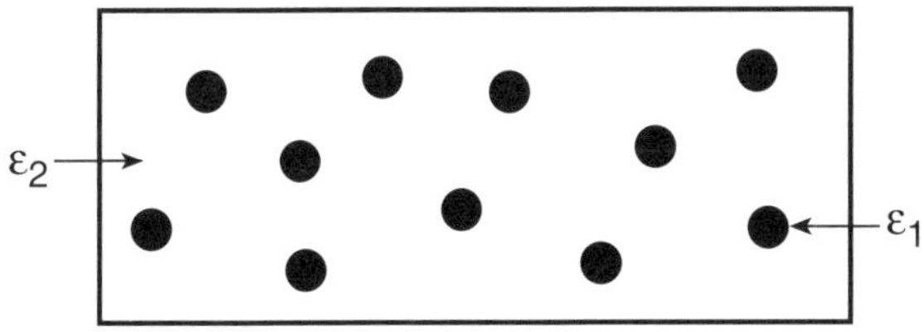

Fig. 2.13. Maxwell–Garnett nanocomposite

The basic idea is that the external light field polarizes the spheres and their own local field is of Lorentzian type. We can apply the Clausius–Mosotti relation, which can be expressed relative to the medium of permittivity ε_2, as follows:

$$\frac{\varepsilon_{\text{eff}}(\omega) - \varepsilon_2(\omega)}{\varepsilon_{\text{eff}}(\omega) + 2\varepsilon_2(\omega)} = \frac{1}{3\varepsilon_2(\omega)} \sum_j \aleph_j \alpha_j(\omega), \tag{2.57}$$

where $\aleph_j$ is the concentration and α_j the microscopic polarizability of the atoms of type j. For a two-phase nanocomposite we can then write

$$\aleph_{1+2}\alpha_{\text{eff}}(\omega) = \aleph_{1+2}\left[\alpha_1(\omega) + \alpha_2(\omega)\right], \tag{2.58}$$

where

$$\aleph_{1+2}\alpha_{\text{eff}} = f_1 \aleph_1 \alpha_1 + (1 - f_1)\aleph_2 \alpha_2. \tag{2.59}$$

Application of the Clausius–Mosotti relation to (2.58), together with (2.59), results in the Maxwell–Garnett formula

$$\frac{\varepsilon_{\text{eff}}(\omega) - \varepsilon_2(\omega)}{\varepsilon_{\text{eff}}(\omega) + 2\varepsilon_2(\omega)} = f_1 \frac{\varepsilon_1(\omega) - \varepsilon_2(\omega)}{\varepsilon_1(\omega) + 2\varepsilon_2(\omega)}. \tag{2.60}$$

Again the effective permittivity is a holomorphic function in the upper half plane and, from the definition of (2.54), we find that the effective complex refractive index is also a holomorphic function in the upper half plane.

As a summary of the two examples, the Kramers–Kronig relation given by

$$\mathrm{i}\pi\left[N_{\text{eff}}(\omega') - 1\right] = \mathrm{P}\int_{-\infty}^{\infty} \frac{N_{\text{eff}}(\omega) - 1}{\omega - \omega'}\,\mathrm{d}\omega \tag{2.61}$$

can be found for two-phase nanocomposites. Owing to possible complex zeros, the phase retrieval procedure of the complex reflectance of nanocomposites, in the general case, is a complicated matter and may be treated using the models of Toll [24].

2.8 Dispersion Relations in Magneto-Optics

One of the standard optical spectral measurements involves the detection of the circular dichroism of natural optical and magneto-optical activity. The measurement system is basically similar to the one in Fig. 2.3 but it uses a device, called a spectropolarimeter, in which an external magnetic field (usually a DC-field) is applied in order to induce circular dichroism and birefringence in the sample. On the microscopic scale, the magnetic field reduces the spatial symmetry of the system, because of the Lorentz force (this force provides a new term in (1.1)), which is a vector product of the electron velocity and the magnetic field. As a result the permittivity tensor contains asymmetric elements. Theoretically, the simplest cases occur when the light is propagating either parallel or perpendicular to the applied magnetic field. The most typical measurement configuration is that of Faraday, i.e. light propagation parallel to the magnetic field. The other configuration, where the light propagation is perpendicular to the magnetic field, is called the Voigt configuration.

We assume that spatial dispersion is not present and study the circular dichroism and birefringence of nonmagnetic, isotropic insulators. Furthermore it is assumed that the incident light can be expressed by $e^{i(\boldsymbol{\kappa}\cdot\boldsymbol{r}-\omega t)}$, where $\boldsymbol{\kappa}$ is the wave vector. The light propagation can be then characterized with the aid of Maxwell's equations, (1.3) and the wave vector in the following way:

$$\overline{\boldsymbol{\kappa}} \times \boldsymbol{\kappa} \times \boldsymbol{E} + \frac{\omega^2}{c^2}\widehat{\varepsilon} : \boldsymbol{E} = 0. \tag{2.62}$$

In general the permittivity tensor $\widehat{\varepsilon}$ implies a relatively complicated set of equations for the spatial components of (2.62). Next we assume that the permittivity tensor obeys C_3 symmetry. In such a case it can be expressed as

$$\widehat{\varepsilon} = \begin{pmatrix} \varepsilon_{xx} & \varepsilon_{xy} & 0 \\ -\varepsilon_{xy} & \varepsilon_{xx} & 0 \\ 0 & 0 & \varepsilon_{zz} \end{pmatrix}. \tag{2.63}$$

Then we obtain the following from (2.62):

$$\left(\kappa^2 - \kappa_x^2 - (\omega^2/c^2)\varepsilon_{xx}\right) E_x - \left(\kappa_x\kappa_y + (\omega^2/c^2)\varepsilon_{xy}\right) E_y - \kappa_x\kappa_z E_z = 0,$$

$$-\left(\kappa_y\kappa_x - (\omega^2/c^2)\varepsilon_{xy}\right) E_x + \left(\kappa^2 - \kappa_y^2 - (\omega^2/c^2)\varepsilon_{xx}\right) E_y - \kappa_y\kappa_z E_z = 0,$$

$$-\kappa_z\kappa_x E_x - \kappa_z\kappa_y E_y + \left(\kappa^2 - \kappa_z^2 - (\omega^2/c^2)\varepsilon_{zz}\right) E_z = 0. \tag{2.64}$$

In particular, in the Faraday configuration, where the light is assumed to propagate in the z-direction ($\kappa_x = 0$, $\kappa_y = 0$, $\kappa_z \neq 0$), we have

$$\left(\kappa_z^2 - (\omega^2/c^2)\varepsilon_{xx}\right) E_x - (\omega^2/c^2)\varepsilon_{xy} E_y = 0,$$

$$(\omega^2/c^2)\varepsilon_{xy}E_x + \left(\kappa_z^2 - (\omega^2/c^2)\varepsilon_{xx}\right)E_y = 0,$$

$$-(\omega^2/c^2)\varepsilon_{zz}E_z = 0. \tag{2.65}$$

We readily can deduce that $E_z = 0$, and in addition

$$\kappa_z^2(\omega) = \frac{\omega^2}{c^2}\left[\varepsilon_{xx}(\omega) \pm \mathrm{i}\varepsilon_{xy}(\omega)\right]. \tag{2.66}$$

In the case of the Lorentz oscillator [53],

$$\varepsilon_{xx}(\omega) = 1 + \frac{\omega_{\mathrm{p}}^2\left(\omega_0^2 - \omega^2 - \mathrm{i}\Gamma\omega\right)}{\left(\omega_0^2 - \omega^2 - \mathrm{i}\Gamma\omega\right)^2 - \omega^2\omega_{\mathrm{c}}^2},$$

$$\varepsilon_{xy}(\omega) = \frac{-\mathrm{i}\omega_{\mathrm{p}}^2\omega_{\mathrm{c}}\omega}{\left(\omega_0^2 - \omega^2 - \mathrm{i}\Gamma\omega\right)^2 - \omega^2\omega_{\mathrm{c}}^2}, \tag{2.67}$$

where $\omega_{\mathrm{c}} = eH/mc$ is the cyclotron frequency and $\omega_{\mathrm{p}}^2 = \aleph e^2/m\varepsilon_0$ is the plasma frequency. Next we define a complex refractive index

$$N_\pm^2(\omega) = \varepsilon_{xx}(\omega) \pm \mathrm{i}\varepsilon_{xy}(\omega). \tag{2.68}$$

We have the result that $E_y/E_x = \pm\,\mathrm{i}$. In other words only left- and right-handed circularly polarized light modes are allowed in the light propagation [53]. Furthermore, the absorption and dispersion of the light depend on its handedness. This in turn means that the symmetry relation obeyed by the optical constants has a lower symmetry than in the conventional case, as can be deduced from (2.68). Now the real refractive index and the extinction coefficient do not separately obey Kramers–Kronig dispersion relations. The handedness means that the optical constants involve one handedness for positive angular frequencies, and the opposite handedness for negative frequencies [53]. This means also a discontinuity at zero angular frequency in the slope of such optical constants. The functional dependences of $\mathrm{Re}\left\{N_\pm^2(\omega)\right\}$ and $\mathrm{Im}\left\{N_\pm^2(\omega)\right\}$ on ω are illustrated in Figs 2.14 and 2.15 using the Lorentzian model of (2.67).

The symmetry relation for the complex refractive index for circularly polarized modes is now

$$N_\pm(-\hat{\omega}^*) = N_\mp^*(\hat{\omega}), \tag{2.69}$$

where "+" stands for the left- and "−" for the right-handed polarization mode. Now it turns out that the complex functions defined by $1/2\left[n_+(\hat{\omega}) + n_-(\hat{\omega})\right] - 1 + \mathrm{i}/2\left[k_+(\hat{\omega}) + k_-(\hat{\omega})\right]$ and $\left[n_+(\hat{\omega}) - n_-(\hat{\omega})\right] + \mathrm{i}\left[k_+(\hat{\omega}) - k_-(\hat{\omega})\right]$ are holomorphic in the upper half plane and possess the appropriate asymptotic behavior. It is then possible to apply complex contour integration, by introducing a pole on the real axis, and to obtain dispersion relations that resemble the conventional Kramers–Kronig relations:

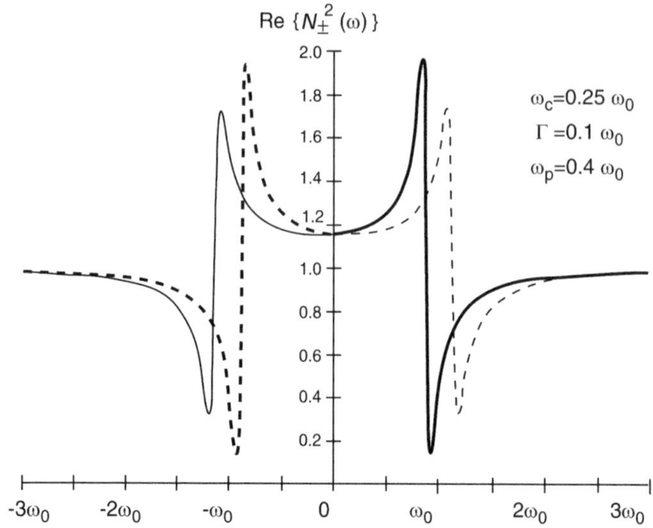

Fig. 2.14. The real parts $\mathrm{Re}\left\{N_\pm^2(\omega)\right\}$, in the Lorentzian model. The real part for the left-handed mode is drawn with *thick lines* and follows the *solid curve* for $\omega > 0$, and the *dashed curve* for $\omega < 0$, so that there is a cusp at the origin. Similar considerations apply to the right-handed mode, indicated by the *thin lines* [53]

$$\frac{1}{2}\left[n_+(\omega') + n_-(\omega')\right] - 1$$

$$= \frac{2}{\pi}\mathrm{P}\int_0^\infty \frac{\omega}{\omega^2 - \omega'^2}\left[\frac{k_+(\omega) + k_-(\omega)}{2}\right]\mathrm{d}\omega$$

$$\frac{1}{2}\left[k_+(\omega') + k_-(\omega')\right]$$

$$= -\frac{2\omega'}{\pi}\mathrm{P}\int_0^\infty \frac{1}{\omega^2 - \omega'^2}\left[\frac{n_+(\omega) + n_-(\omega)}{2} - 1\right]\mathrm{d}\omega. \tag{2.70}$$

We also obtain relations that are different from the Kramers–Kronig relations:

$$n_+(\omega') - n_-(\omega') = \frac{2\omega'}{\pi}\mathrm{P}\int_0^\infty \frac{1}{\omega^2 - \omega'^2}\left[k_+(\omega) - k_-(\omega)\right]\mathrm{d}\omega,$$

$$k_+(\omega') - k_-(\omega') = -\frac{2}{\pi}\mathrm{P}\int_0^\infty \frac{\omega}{\omega^2 - \omega'^2}\left[n_+(\omega) - n_-(\omega)\right]\mathrm{d}\omega. \tag{2.71}$$

The difference $k_+ - k_-$ is known as the magnetic circular dichroism and the difference $n_+ - n_-$ as the magnetic birefringence. Equations (2.70) and (2.71)

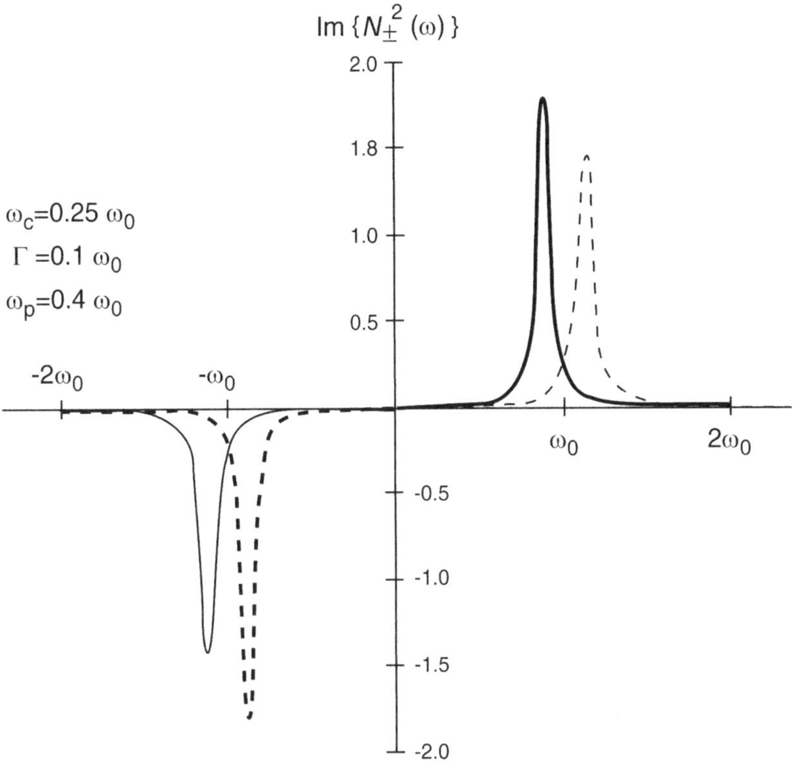

Fig. 2.15. The imaginary parts $\mathrm{Im}\left\{N_{\pm}^{2}(\omega)\right\}$, in the Lorentzian model. The imaginary part for the left-handed mode is drawn with *thick lines* and follows the *solid curve* for $\omega > 0$ and the *dashed curve* for $\omega < 0$. The right-handed mode is indicated by the *thin lines* [53]

hold for isotropic insulators under the influence of a uniform magnetic field. The rotation of the polarization angle can be found from

$$\Phi_{\mathrm{F}}(\omega) = \frac{\omega d}{2c}\left[n_{+}(\omega) - n_{-}(\omega)\right], \tag{2.72}$$

where d is the thickness of the sample.

Dispersion relations for the complex magnetoreflectance [54] can be obtained in a similar manner to that above, taking into account the symmetry relation

$$r_{\pm}\left(\hat{\omega}^{*}\right) = r_{\mp}^{*}\left(\widehat{\omega}\right). \tag{2.73}$$

Kramers–Kronig relations can be constructed as follows,

$$\ln\frac{|r_+(\omega')|}{|r_+(\omega'')|}+\ln\frac{|r_-(\omega')|}{|r_-(\omega'')|}$$

$$=\frac{2}{\pi}\mathrm{P}\int_0^\infty \omega\left[\varphi_+(\omega)+\varphi_-(\omega)\right]\left(\frac{1}{\omega^2-\omega'^2}-\frac{1}{\omega^2-\omega''^2}\right)\mathrm{d}\omega,$$

$$\varphi_+(\omega')+\varphi_-(\omega')=-\frac{2\omega'}{\pi}\mathrm{P}\int_0^\infty \frac{\ln|r_+(\omega)|+\ln|r_-(\omega)|}{\omega^2-\omega'^2}\mathrm{d}\omega, \tag{2.74}$$

and also non-Kramers–Kronig-type dispersion relations,

$$\ln|r_+(\omega')|-\ln|r_-(\omega')|=\frac{2\omega'}{\pi}\mathrm{P}\int_0^\infty \frac{\varphi_+(\omega)-\varphi_-(\omega)}{\omega^2-\omega'^2}\mathrm{d}\omega,$$

$$\varphi_+(\omega')-\varphi_-(\omega')=-\frac{2}{\pi}\mathrm{P}\int_0^\infty \frac{\omega\ln\left|\frac{r_+(\omega)}{r_-(\omega)}\right|}{\omega^2-\omega'^2}\mathrm{d}\omega. \tag{2.75}$$

Finally, without going into details, we remark that in the Voigt configuration normal Kramers–Kronig relations can be constructed for the absorption–dispersion effects. Readers interested in magneto-optical spectroscopy are recommended to read the pedagogically good paper by Smith [53].

3. Dispersion Relations in Nonlinear Optics

The invention of lasers has resulted in novel types of optical spectroscopies for materials research. The basic idea of these spectroscopies is to detect signals that are due to nonlinear optical processes induced by the application of intense laser beams to the material. The most popular type of nonlinear optical spectroscopy makes use of two laser beams, where one beam is for pumping and the other for probing. The wavelength of the pumping laser is usually fixed, whereas the wavelength of the probe beam is tuned. Such a technique is exploited, for example, in coherent anti-Stokes Raman spectroscopy (CARS) [55].

The common feature of the various nonlinear processes that can occur in isotropic or anisotropic materials is that the interaction of the light with the material is governed by the microscopic properties of the material. The macroscopic nonlinear response of a nonlinear optical material can be characterized by the electric charge polarization, which now manifests the nonlinearity of the material via a nonlinear susceptibility. Nonlinear susceptibilities are complex functions of combinations of the angular frequencies of the fields propagating in the nonlinear material.

We start the description of nonlinear susceptibilities using classical models from mechanics and electromagnetism. Just as in the regime of linear optics, we consider here the concept of causality and the existence of dispersion relations for the nonlinear susceptibilities. Finally we find out that the properties of the nonlinear susceptibilities are more complicated than those of the linear optical constants, and it is possible that conventional dispersion relations cannot be found for particular degenerate nonlinear susceptibilities.

3.1 Hyperpolarizability

Maxwell predicted the possibility that the interaction of an electromagnetic wave with a material can depend on the strength of the incident field. The nonlinear nature of the magnetization of ferromagnetic materials has been applied for a long time in technology. The light field induced by the nonlinear optical response of a material could be detected by experiments immediately after the invention of the laser. The electric polarization can be affected by the strength of the external electric field. In the theory of electromagnetism we usually make the dipole approximation, which means that the electric polarization can be given by a power series expansion

$$P(E) = P(0) + \left(\frac{\partial P}{\partial E}\right)_0 E + \frac{1}{2!}\left(\frac{\partial^2 P}{\partial E^2}\right)_0 E^2 + \cdots, \tag{3.1}$$

where $E = E_0 \exp(-\mathrm{i}\omega t)$ is the electric field. For nonpolar materials the first term on the right-hand side in (3.1) is zero and we can rewrite the contents of (3.1) by introducing the concept of nonlinear susceptibility:

$$P(E) = \chi^{(1)}E + \chi^{(2)}E^2 + \chi^{(3)}E^3 + \cdots. \tag{3.2}$$

The first term is nothing but the result from linear response theory, whereas the higher terms in the series expansion describe the so-called hyperpolarizabilities. The hyperpolarizabilities in turn are given by the nonlinear susceptibilities $\chi^{(n)}$, where n is the order. We observe from (3.2) that in this series expansion harmonic frequencies $n\omega$, $n = 1, 2, ...$, appear. Importantly, powers of the amplitude of the electric field have a key role in the strength of the components of the hyperpolarizabilities. This is a fact that can be exploited, especially when using lasers as light sources in order to induce nonlinear optical phenomena.

In reality, the nonlinear susceptibilities are tensors that relate components of the nonlinear polarization vector to various components of the optical field vectors. In addition, the nonlinear susceptibilities have the spatial symmetry properties of the particular medium. We are interested mainly in the elements of the nonlinear susceptibility tensor, which are considered as scalar complex quantities that depend on the angular frequencies. The most important nonlinear optical processes are related to the second-order (for anisotropic materials) or third-order (both isotropic and anisotropic materials) nonlinearities, i.e. processes characterized by the material properties $\chi^{(2)}$ and $\chi^{(3)}$.

Consider next the usual case of parallel pump and probe waves, where two electric fields, with different angular frequencies ω_1 and ω_2, so that $E = E_1 \exp(-\mathrm{i}\omega_1 t) + E_2 \exp(-\mathrm{i}\omega_2 t)$, are incident on a nonlinear material. Then the total polarization, which is the sum of the linear and nonlinear polarizations, is found to be

$$\begin{aligned} P = {} & \chi^{(1)}E_1 \mathrm{e}^{-\mathrm{i}\omega_1 t} + \chi^{(1)}E_2 \mathrm{e}^{-\mathrm{i}\omega_2 t} \\ & + \chi^{(2)}E_1^2 \mathrm{e}^{-2\mathrm{i}\omega_1 t} + \chi^{(2)}E_2^2 \mathrm{e}^{-2\mathrm{i}\omega_2 t} \\ & + 2\chi^{(2)}E_1 E_2 \mathrm{e}^{-\mathrm{i}(\omega_1+\omega_2)t} \\ & + \chi^{(3)}E_1^3 \mathrm{e}^{-3\mathrm{i}\omega_1 t} + 3\chi^{(3)}E_1^2 E_2 \mathrm{e}^{-\mathrm{i}(2\omega_1+\omega_2)t} \\ & + 3\chi^{(3)}E_1 E_2^2 \mathrm{e}^{-\mathrm{i}(\omega_1+2\omega_2)t} + \chi^{(3)}E_2^3 \mathrm{e}^{-3\mathrm{i}\omega_2 t} + \cdots. \end{aligned} \tag{3.3}$$

Now we can distinguish in (3.3) waves oscillating at the fundamental, harmonic and sum angular frequencies. We prefer the notation of complex electric fields; however, remembering that actually the electric fields have to be real, we apply of Euler's formulae to the complex electric fields:

$$E_1 \cos(\omega_1 t) = (E_1/2)\left(\mathrm{e}^{\mathrm{i}\omega_1 t} + \mathrm{e}^{-\mathrm{i}\omega_1 t}\right),$$
$$E_2 \cos(\omega_2 t) = (E_2/2)\left(\mathrm{e}^{\mathrm{i}\omega_2 t} + \mathrm{e}^{-\mathrm{i}\omega_2 t}\right). \quad (3.4)$$

We then find that in the expansion of (3.3) terms describing difference frequency generation also appear. In the pump–probe scheme frequency combinations such as ω_1, ω_2, $3\omega_1$, $3\omega_2$, $|2\omega_1 \pm \omega_2|$ and $|2\omega_2 \pm \omega_1|$ may occur in polarization because of the third-order nonlinearity.

The generation of the new frequencies opens up the rich world of nonlinear processes, which provide, among other technologies, new types of optical spectroscopies.

3.2 Anharmonic Lorentz Oscillator Model

Next we approach the problem of the nonlinearity of media by considering the equation of motion of an electron. This can be accomplished by inspection of the classical Lorentz oscillator, but now introducing anharmonic terms into the equation of motion. We can generalize (1.1)

$$\frac{\mathrm{d}^2 x}{\mathrm{d}t^2} + \Gamma \frac{\mathrm{d}x}{\mathrm{d}t} + \omega_0^2 x + \sum_{j=2}^{\infty} \beta_j x^j = -\frac{e}{m} E_0 \mathrm{e}^{-\mathrm{i}\omega t}, \quad (3.5)$$

where the sum $\sum \beta_j \chi^j$ denotes the anharmonic terms, which can be considered as extra driving forces if they are transferred to the right-hand side of (3.5). The introduction of the anharmonic terms has changed the original second-order linear differential equation to a nonlinear one, whose solution is usually far from trivial. However, mathematical methods to solve nonlinear differential equations have been established, especially in the context of solitary wave propagation [56]. Here we seek a solution by applying a series expansion

$$x = \sum_{l=1}^{\infty} x_l E^l. \quad (3.6)$$

Then the polarization, including the linear and nonlinear contributions, can be expressed as

$$P = -\aleph e \sum_{l=1}^{\infty} x_l E^l. \quad (3.7)$$

For the sake of simplicity we take only the two first terms in the sum of the anharmonic terms in (3.5). After substitution of the trial function (3.6) into (3.5) we obtain a set of equations with the same power of the electric field E on both sides of the equations:

$$
\begin{aligned}
\frac{\mathrm{d}^2 x_1}{\mathrm{d}t^2} + \Gamma \frac{\mathrm{d}x_1}{\mathrm{d}t} + \omega_0^2 x_1 &= -\frac{e}{m} E_0 \mathrm{e}^{-\mathrm{i}\omega t}, \\
\frac{\mathrm{d}^2 x_2}{\mathrm{d}t^2} + \Gamma \frac{\mathrm{d}x_2}{\mathrm{d}t} + \omega_0^2 x_2 &= \beta_2 x_1^2, \\
\frac{\mathrm{d}^2 x_3}{\mathrm{d}t^2} + \Gamma \frac{\mathrm{d}x_3}{\mathrm{d}t} + \omega_0^2 x_3 &= 2\beta_2 x_1 x_2, + \beta_3 x_1^3 \\
\vdots \qquad \vdots \qquad & \quad \vdots \qquad \vdots \; .
\end{aligned} \tag{3.8}
$$

The first equation is already familiar from the case of the linear Lorentz susceptibility, and its solution was given in (1.2). Next we use the solution for x_1 and substitute it into the second differential equation in (3.8), which is now a conventional second-order differential equation, and solve this equation. The procedure is continued by substituting the solutions x_1 and x_2 into the third differential equation in (3.8), whereafter we solve that equation, etc. The solutions are

$$
\begin{aligned}
x_1 &= -\frac{e}{m} \frac{E_0 \mathrm{e}^{-\mathrm{i}\omega t}}{\omega_0^2 - \omega^2 - \mathrm{i}\Gamma\omega}, \\
x_2 &= \frac{e^2}{m^2} \beta_2 \frac{E_0^2 \mathrm{e}^{-2\mathrm{i}\omega t}}{\left(\omega_0^2 - 4\omega^2 - 2\mathrm{i}\Gamma\omega\right)\left(\omega_0^2 - \omega^2 - \mathrm{i}\Gamma\omega\right)^2}, \\
x_3 &= -\frac{e^3}{m^3} \left[\frac{2\beta_2^2}{\omega_0^2 - 4\omega^2 - 2\mathrm{i}\Gamma\omega} + \beta_3 \right] \\
&\quad \times \frac{E_0^3 \mathrm{e}^{-3\mathrm{i}\omega t}}{\left(\omega_0^2 - \omega^2 - \mathrm{i}\Gamma\omega\right)^3 \left(\omega_0^2 - 9\omega^2 - 3\mathrm{i}\Gamma\omega\right)}, \\
&\quad \vdots \qquad \qquad \vdots
\end{aligned} \tag{3.9}
$$

where the last two expressions describe the generation of harmonic waves.

In the case of electric fields having three different angular frequencies ω_1, ω_2 and ω_3 acting as the driving forces on the right-hand side of (3.5), we finally find, using the definitions in (3.2) and (3.7) the desired results that enable us to describe the complex second- and third-order susceptibilities as functions of the angular frequencies:

$$
\begin{aligned}
\chi^{(2)}\left(\omega_1, \omega_2\right) &= -\aleph \beta_2 \frac{e^3}{m^2} F\left(\omega_1\right) F\left(\omega_2\right) F\left(\omega_1 + \omega_2\right), \\
\chi^{(3)}\left(\omega_1, \omega_2, \omega_3\right) & \\
&= \aleph \frac{e^4}{m^3} \left[\beta_3 + \frac{2\beta_2^2}{3} \left[F\left(\omega_1 + \omega_2\right) + F\left(\omega_2 + \omega_3\right) + F\left(\omega_1 + \omega_3\right) \right] \right] \\
&\quad \times F\left(\omega_1\right) F\left(\omega_2\right) F\left(\omega_3\right) F\left(\omega_1 + \omega_2 + \omega_3\right),
\end{aligned} \tag{3.10}
$$

where

$$F(\omega) = \frac{1}{\omega_0^2 - \omega^2 - \mathrm{i}\Gamma\omega}. \tag{3.11}$$

The functional dependences of the real and imaginary parts of the so-called degenerate complex nonlinear susceptibilities $\chi^{(2)}(\omega,\omega)$, $\chi^{(2)}(\omega,-\omega)$, $\chi^{(3)}(\omega,\omega,\omega)$ and $\chi^{(3)}(\omega,\omega,-\omega)$, given with the aid of (3.10) and (3.11), are illustrated in Fig. 3.1. Note carefully that the second- and third-order nonlinear susceptibilities are holomorphic functions in the upper half planes corresponding to the complex variables $\widehat{\omega}_1$, $\widehat{\omega}_2$ and $\widehat{\omega}_3$, provided that there is a plus sign in front of each angular frequency. Indeed, we can generalize the concept of holomorphism to deal with analysis of a function of several complex variables [57] (a brief introduction of the theory of several complex variables is presented in Appendix E). On the other hand, when there appears a minus sign in front of one angular frequency while the others possess a plus sign, then we can find poles in the upper half plane corresponding to the angular frequency with a minus sign. However, such a function can be considered to be holomorphic in the lower half plane with respect to the "pathological" angular frequency variable. The most striking situation appears in the case of a degenerate nonlinear susceptibility such as $\chi^{(3)}(\omega,\omega,-\omega)$. This has poles simultaneously in the lower and upper half planes of the same complex

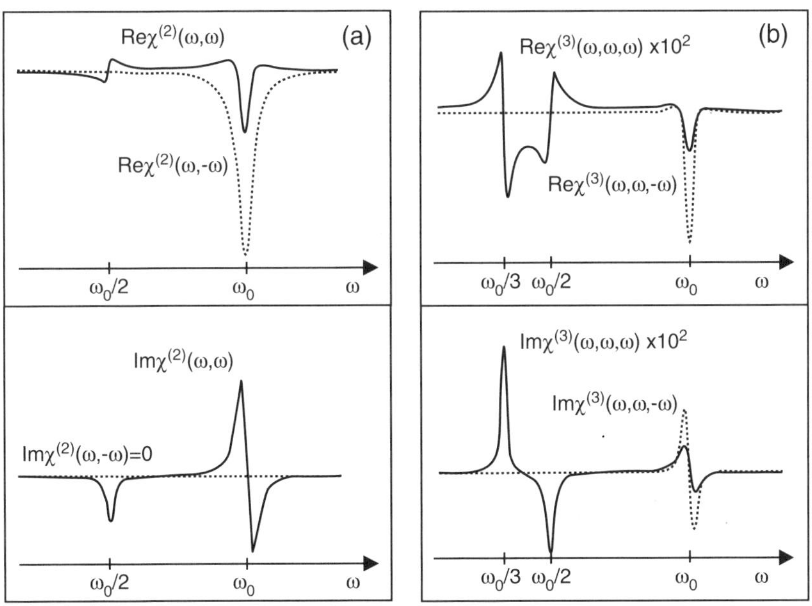

Fig. 3.1. Degenerate nonlinear susceptibilities: **(a)** real and imaginary parts of $\chi^{(2)}$ and **(b)** real and imaginary parts of $\chi^{(3)}$. The *solid lines* correspond to the angular frequency combinations (ω,ω) and (ω,ω,ω), and the *dashed lines* to the combinations $(\omega,-\omega)$ and $(\omega,\omega,-\omega)$

plane. Therefore we can expect problems in the formulation of conventional "Hilbert-type" dispersion relations for such complex functions. This degenerate nonlinear susceptibility is, instead of being holomorphic, a meromorphic function [10, 13]. We spend some time later on the dispersion theory of meromorphic nonlinear susceptibilities.

3.3 Nonlinear Susceptibilities and Causality

Causality is a basic physical principle that holds when we consider the interaction of an electromagnetic field with a material, whether it responds in either a linear or a nonlinear manner. We can write, for a holomorphic nonlinear susceptibility,

$$\chi^{(n)}\left(\omega_1, \omega_2, \cdots, \omega_n\right) = \int_0^\infty \mathrm{d}t_1 \int_0^\infty \mathrm{d}t_2 \cdots \int_0^\infty \mathrm{d}t_n \chi^{(n)}\left(t_1, t_2, \cdots, t_n\right) \mathrm{e}^{\mathrm{i}\left(\omega_1 t_1 + \omega_2 t_2 + \cdots + \omega_n t_n\right)}, \quad (3.12)$$

where the lower bounds of the integrals are limited to zero by the principle of causality. The reality of the time-dependent nonlinear susceptibility in turn implies a symmetry relation

$$\chi^{(n)}\left(-\omega_1, -\omega_2, \cdots, -\omega_n\right) = \left[\chi^{(n)}\left(\omega_1, \omega_2, \cdots, \omega_n\right)\right]^*, \quad (3.13)$$

which can be generalized to complex frequencies just as in the regime of linear optics. In the case of the meromorphic nonlinear susceptibilities that are needed in four-wave-mixing spectroscopy, the time-dependent nonlinear susceptibility or correlation function can be calculated, and (detailed calculations were presented by Kircheva and Hadjichristov [28]) there is no break of causality for such susceptibilities. The conclusion of the investigations of Kircheva and Hadjichristov was that causality is a necessary but not sufficient condition for the existence of conventional dispersion relations. Therefore Kramers–Kronig relations cannot be found for meromorphic nonlinear susceptibilities. The reason is the simultaneous appearance of poles in both half planes in the case of a degenerate meromorphic nonlinear susceptibility. Later we learn that a meromorphic nonlinear susceptibility can possess, in addition to the poles, complex zeros. Then, especially in phase retrieval problems, further complications arise because of the usual way of treating the logarithm of the modulus of the nonlinear susceptibility.

3.4 Dispersion Relations for Holomorphic Nonlinear Susceptibilities

The problem of the existence of Kramers–Kronig dispersion relations for nonlinear responses was investigated in the early sixties by Kogan [58], Price

[59] and Caspers [60], who considered second-order nonlinearities of matter. Ridener and Good [61] derived dispersion relations for third- and arbitrary-order nonlinear responses, whereas Smet and Groenendael [62] showed that no dispersion relations can be established between different nonlinear phenomena. All these studies were performed in such a way that only one angular frequency of the incident electric field was permitted to change, whereas the other (pump) fields were assumed to have fixed frequencies. In addition there was an assumption about the holomorphicity of the nonlinear susceptibility with respect to the variable angular frequency. The derivation of the dispersion relations was based on the complex contour integration of one complex variable, as in the case of the linear optical constants. The nonlinear susceptibilities have an asymptotic fall-off that is similar to or, usually, stronger than that of the linear susceptibility for high angular frequencies. Then it is possible to write a dispersion relation for the nth-order susceptibility as follows:

$$\chi^{(n)}\left(\omega_1,\omega_2,\cdots,\omega_j'\cdots,\omega_n\right)=\frac{1}{\mathrm{i}\pi}\mathrm{P}\int_{-\infty}^{\infty}\frac{\chi^{(n)}\left(\omega_1,\omega_2,\cdots,\omega_j'\cdots,\omega_n\right)}{\omega_j-\omega_j'}\mathrm{d}\omega_j. \tag{3.14}$$

This is the form that usually corresponds to the conventional experimental scheme of using one probe wave, and pump waves of fixed angular frequency.

A more general dispersion relation can be derived with the aid of complex analysis of several complex angular frequency variables, $\widehat{\omega}_j=\omega_j+\mathrm{i}v_j$. In this case we permit the variation of the angular frequencies of all electric fields. Now the multiple domain of complex half planes provides us with information about the region of holomorphicity of $\chi^{(n)}\left(\omega_1,\omega_2,\cdots,\omega_n\right)$. Each complex variable is considered separately while the others are held constant. Application of complex contour integration separately for each variable yields the dispersion relation [63, 64]

$$\begin{aligned}&\chi^{(n)}\left(\omega_1',\omega_2',\cdots,\omega_n'\right)\\&=\frac{1}{(\mathrm{i}\pi)^n}\mathrm{P}\int_{-\infty}^{\infty}\int_{-\infty}^{\infty}\cdots\int_{-\infty}^{\infty}\frac{\chi^{(n)}\left(\omega_1,\omega_2,\cdots,\omega_n\right)}{\left(\omega_1-\omega_1'\right)\left(\omega_2-\omega_2'\right)\left(\cdots\right)\left(\omega_n-\omega_n'\right)}\mathrm{d}\omega_1 d\omega_2\cdots\mathrm{d}\omega_n.\end{aligned} \tag{3.15}$$

Illustration of the the complex contour integration procedure is a complicated matter since we have in the general case n different complex planes. However, if we assume only two fields, with different variable frequencies and a common imaginary axis, then the integration contours can be closed in the corresponding upper half planes as in Fig. 3.2.

We observe from (3.15) that only the odd orders will yield dispersive–dissipative effects.

It took a considerable long time before the validity of the dispersion relations for nonlinear susceptibilities could be experimentally tested. The valid-

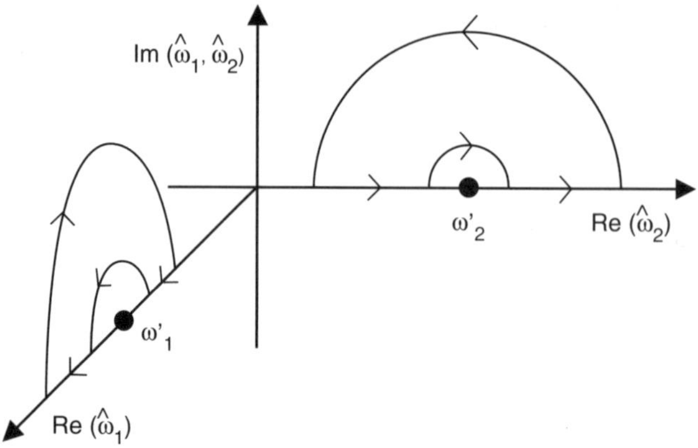

Fig. 3.2. Illustration of contours of integration for several complex angular frequency variables [65]

ity was confirmed by Kishida et al. [66] by tedious ellipsometric experiments using a diversity of laser lines, applied to third-harmonic generation from polysilane films [67]. Polysilane belongs to a group of silicon-based polymers that can be regarded as chemically synthesized quasi-one-dimensional semiconductors. The measurements covered an angular frequency range that included two- and three-photon resonances. The phase retrieval procedure using the Kramers–Kronig relations gave results in accordance with the experiments, as illustrated in Fig. 3.3.

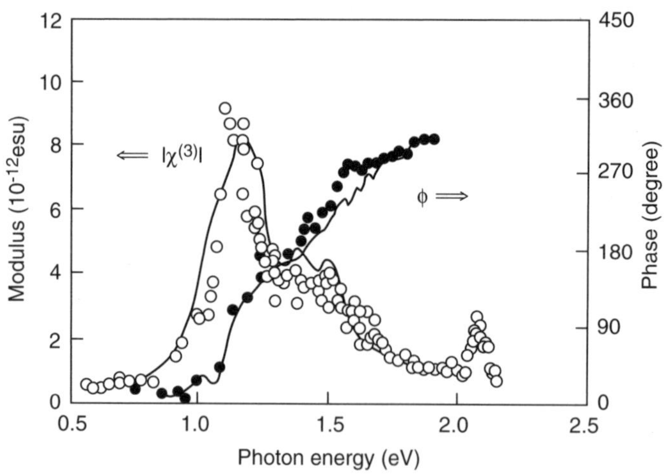

Fig. 3.3. Spectra of the modulus and phase of the complex third-order nonlinear susceptibility of a polydihexylsilane film: *solid lines*, Kramers–Kronig analysis; *circles*, experimental values [66]

The nonlinear susceptibility will induce nonlinear contributions to the real refractive index and the corresponding extinction coefficient. In this case the corresponding complex refractive index is $N = N_{\mathrm{L}} + N_{\mathrm{NL}}$, where N_{L} is the linear and N_{NL} is the nonlinear contribution. Now in the conventional system of one probe and one pump beam, the complex refractive index is a holomorphic function in the upper half plane ($\mathrm{Im}\,\widehat{\omega}_1 \geq 0$) and ω_2 is fixed. The complex refractive index can be written as

$$\begin{aligned} N\left(\widehat{\omega}_1, \omega_2, E_2\right) &= N_{\mathrm{L}}\left(\widehat{\omega}_1\right) + N_{\mathrm{NL}}\left(\widehat{\omega}_1, \omega_2; E_2\right) \\ &= N_{\mathrm{L}}\left(\widehat{\omega}_1\right) + N_{\mathrm{NL},1} E_2^2 + \cdots , \end{aligned} \tag{3.16}$$

where E_2 is the electric field of the pump beam. Then, according to Bassani and Scandolo [68], we can write the Kramers–Kronig relations

$$\begin{aligned} n\left(\omega_1', \omega_2; E_2\right) &= \frac{2}{\pi} \mathrm{P} \int_0^\infty \frac{\omega_1 k\left(\omega_1, \omega_2; E_2\right)}{\omega_1^2 - \omega_1'^2} \mathrm{d}\omega_1, \\ k\left(\omega_1', \omega_2; E_2\right) &= \frac{-2\omega_1}{\pi} \mathrm{P} \int_0^\infty \frac{n\left(\omega_1, \omega_2; E_2\right) - 1}{\omega_1^2 - \omega_1'^2} \mathrm{d}\omega_1 . \end{aligned} \tag{3.17}$$

Note that in this case the static real refractive index depends on the pump wave, i.e.

$$n\left(0, \omega_2; E_2\right) = \frac{2}{\pi} \mathrm{P} \int_0^\infty \frac{k\left(\omega_1, \omega_2; E_2\right)}{\omega_1} \mathrm{d}\omega_1 . \tag{3.18}$$

3.5 Examples of Meromorphic Nonlinear and Total Susceptibilities

Before going into the details of the general dispersion theory of meromorphic nonlinear susceptibilities, we first study two examples. One of them is related to the third-order meromorphic nonlinear susceptibility, which is obtained with the aid of (3.10) as follows:

$$\chi^{(3)}\left(\omega, \omega, -\omega,\right) = \aleph \frac{e^4}{m^3} \left[\beta_3 + \frac{2\beta_2^2}{3} \left[F\left(2\omega\right) + 2F\left(0\right)\right]\right] F^3\left(\omega\right) F\left(-\omega\right) . \tag{3.19}$$

Now we observe immediately that the degenerate nonlinear susceptibility possesses simultaneously poles in the lower and upper half planes. The poles in the complex angular frequency plane can be found by solving the two algebraic equations

$$\omega_0^2 - \widehat{\omega}^2 - \mathrm{i}\Gamma\widehat{\omega} = 0,$$

$$\omega_0^2 - \widehat{\omega}^2 + \mathrm{i}\Gamma\widehat{\omega} = 0. \tag{3.20}$$

Note that the poles in the lower half plane are of third order. Most often, in optical spectrum analysis, we deal with a meromorphic function given by a quotient $f(z)/g(z)$, where f and g are holomorphic functions.

The functional behavior of the real and imaginary parts of the meromorphic nonlinear susceptibility, which is usually related to degenerate four-wave-mixing processes, can be seen in Fig. 3.1.

This meromorphism is not limited to nonlinear susceptibilities. As an example we mention the meromorphic total susceptibility of a two-level atomic system given by Yariv [69]. The theoretical model of Yariv for the total susceptibility is based on the density matrix model and its steady-state solution. The detailed derivation of the total susceptibility can be found in Yariv's book; here we state only the result, as follows:

$$\chi(\omega) = -\frac{\mu^2 \triangle\aleph_0}{\varepsilon_0 \hbar} \frac{\omega_0 - \omega + \mathrm{i}/T_2}{(\omega - \omega_0)^2 + (1 + I/I_\mathrm{s})/T_2^2}, \tag{3.21}$$

where μ is the dipole moment for the transition frequency ω_0, which has a dephasing time T_2 and a saturation irradiance I_s and $\triangle\aleph_0$ is the ground-state population density. The modulus and real and imaginary parts of the total susceptibility given by (3.21) are illustrated in Fig. 3.4.

By inspection of the total susceptibility we observe that there are two poles, located in both half planes, and in addition one zero located in the upper half plane, given by

$$\omega_\mathrm{poles} = \omega_0 \pm (\mathrm{i}/T_2)(1 + I/I_\mathrm{s})^{1/2},$$

$$\omega_\mathrm{zero} = \omega_0 + \mathrm{i}/T_2. \tag{3.22}$$

Certainly, this type of function cannot be characterized as a holomorphic function in the whole upper half plane. Therefore the assumptions made in the derivation of the Kramers–Kronig relations do not hold, and this also implies the invalidity of the Kramers–Kronig relations [28, 63, 64, 69–71] Note carefully that most often, in nonlinear optical spectroscopy, we gain information about the modulus of the nonlinear susceptibility. In particular, when using a logarithm of the modulus of the total susceptibility of the two-level atom, as is usually needed in phase retrieval procedures, the complex zero in the upper half plane is a strong singular point. However, in the general theory of meromorphic functions [13], complex zeros are allowed. The determination of the existence of complex zeros using measured spectra may be problematic [27]. One way to describe complex zeros theoretically is based on the use of the Blaschke product [10, 24], provided that there is a symmetry among the

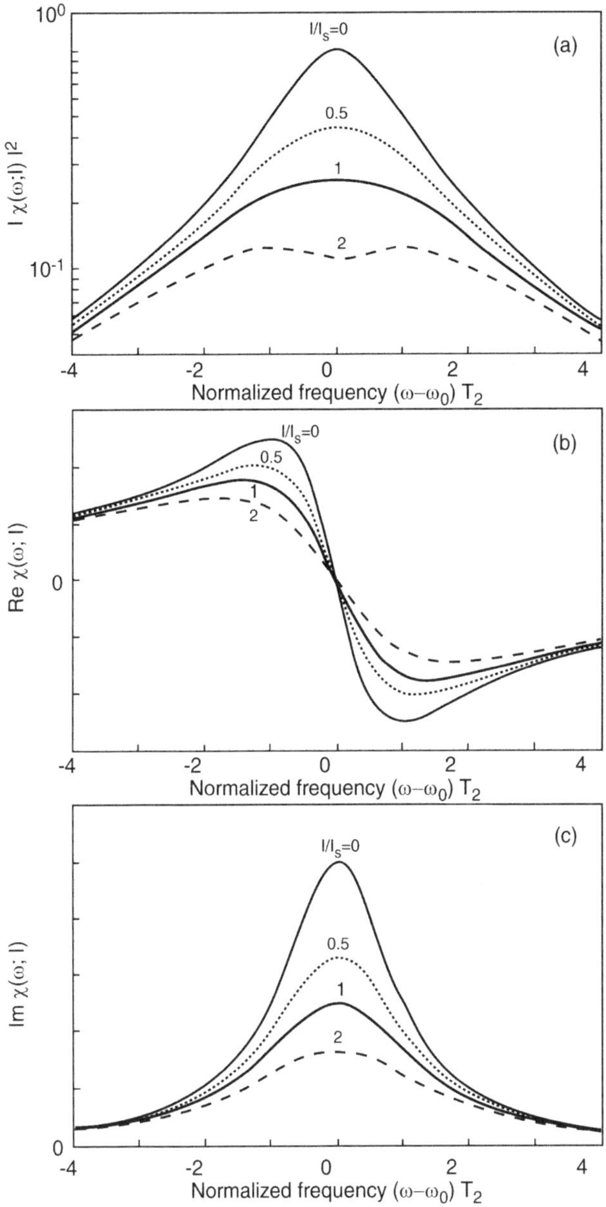

Fig. 3.4. Spectra of total susceptibility χ given by (3.21): (**a**) $|\chi|^2$, (**b**) $\mathrm{Re}\{\chi\}$ and (**c**) $\mathrm{Im}\{\chi\}$

zeros and poles. Then an optical quantity can be described as a meromorphic function in the whole complex plane. However, the total susceptibility considered above does not posses such a symmetry among poles and zeros, since $I \neq 0$. In addition it is already meromorphic in the upper half plane. We solve the problem of the estimation of the existence of zeros in the course of considering sum rules for meromorphic susceptibilities in Chap. 6.

We emphasize once again that there is no problem with the principle of causality, which is always valid, but causality does not imply the existence of Kramers–Kronig relations, as shown by Kircheva and Hadjichristov [28].

Finally we remark that in the case of femtosecond nonlinear optical spectroscopy, that is to say in time-resolved spectroscopy using short pump and probe light pulses, the Kramers–Kronig relations are not always valid [72–74]. If the pump beam is incident on the material before the probe arrives (the system has been disturbed before the probe pulse) then a minus sign appears in the the Kramers–Kronig relations instead of the conventional plus sign. In the case of zero delay, the Kramers–Kronig relations do not hold, because of the simultaneous appearance of poles in both half planes [72–74]. This is nothing but a meromorphic-function property of the nonlinear susceptibility.

3.6 Dispersion Theory of Meromorphic Susceptibilities

The classical dispersion theory of the linear susceptibility by Lorentz and Drude, as well as the quantum mechanical description by Kramers and Heisenberg, allows the number of poles to be countably infinite. Also, the linear susceptibility diverges at the poles. We may state that such a function behaves like a holomorphic function almost everywhere. If, however, we consider the linear susceptibility in the whole complex plane, then we must call it meromorphic. Fortunately, in the derivation of the Kramers–Kronig relations for normal cases, we can avoid the lower half plane by dealing with a function that is holomorphic in the other half plane. According to the principle of causality, such a procedure is also inevitable, if considered in the context of (2.14). The situation with nonlinear susceptibilities is very similar to this. However, nonlinear processes that demand the simultaneous existence of poles in both half planes require the generalization of the treatment of the nonlinear susceptibility, because it is also a meromorphic function in the upper half plane. A meromorphic function can be decomposed into holomorphic and principal parts [75, 76]. Then the meromorphic third-order nonlinear susceptibility has the representation

$$\chi^{(3)}(\widehat{\omega},\widehat{\omega},-\widehat{\omega}) = f(\widehat{\omega}) + \sum_j \Big[c_{1j}(\widehat{\omega}-\widehat{\omega}_{\mathrm{p}j})^{-1} + c_{2j}(\widehat{\omega}-\widehat{\omega}_{\mathrm{p}j})^{-2} + \cdots + c_{lj}(\widehat{\omega}-\widehat{\omega}_{\mathrm{p}j})^{-l} \Big], \tag{3.23}$$

where f represents the holomorphic part and the series expansion represents the principal part of the nonlinear susceptibility. The c_{lj}s are complex numbers (constants), $\widehat{\omega}_{\mathrm{p}j}$ denotes the poles that are located in the upper and lower half planes, and l is the order of the pole.

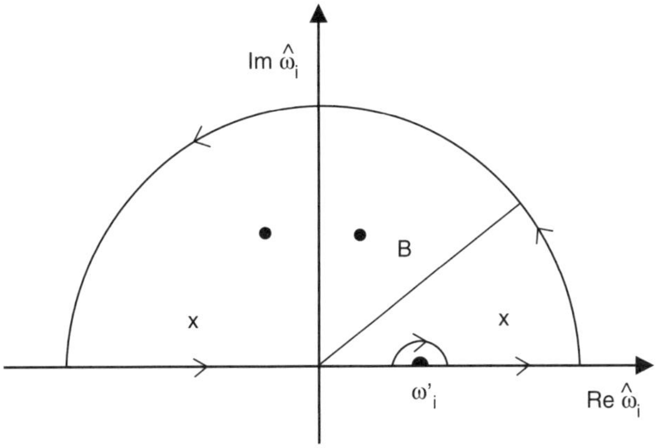

Fig. 3.5. Contour enclosing poles (*crosses*) and zeros (*filled circles*) of meromorphic susceptibility

As a curiosity, if we consider the classical linear Lorentz susceptibility, then $f = 0$. If in turn we are interested in the permittivity then $f = 1$, provided that these functions are considered in the whole complex plane. Fortunately, the function f is often rather simple regardless of whether we are describing linear or nonlinear optical processes.

We proceed to discuss the existence of a dispersion relation for a meromorphic susceptibility. We can indeed find a dispersion relation, which can be called a modified dispersion relation if it is compared with the complex form of the Hilbert transform. Now, recognizing the appearance of poles in the upper half plane and the appropriate fall-off of $\chi^{(3)}(\omega, \omega, -\omega)$ as ω tends to infinity, we can apply complex contour integration, as shown in Fig. 3.5.

Once again the concept of the Cauchy principal value, Jordan's lemma and the theorem of residues can be exploited. Although the most interesting nonlinear susceptibility for the purposes of technology seems to be $\chi^{(3)}(\omega, \omega, -\omega)$, we give here the modified dispersion relation for the nth-order nonlinear susceptibility in the case where all angular frequencies are constants except for the pair $(\omega_j, -\omega_j)$, which introduces the meromorphism. Then we have [77, 78]

$$\begin{aligned} &\mathrm{P}\int_{-\infty}^{\infty} \frac{\chi^{(n)}(\omega_1, \omega_2, \ldots, \omega_j, -\omega_j, \ldots, \omega_n)}{\omega_j - \omega_j'} \mathrm{d}\omega_j \\ &\quad = \mathrm{i}\pi\chi^{(n)}(\omega_1, \omega_2, \ldots, \omega_j', -\omega_j', \ldots \omega_n) \\ &\quad\quad + 2\pi\mathrm{i} \sum_{\substack{\text{poles} \\ \mathrm{Im}\, \widehat{\omega}_j \geq 0}} \mathrm{Res}\left[\frac{\chi^{(n)}(\omega_1, \omega_2, \ldots, \widehat{\omega}_j, -\widehat{\omega}_j, \ldots \omega_n)}{\widehat{\omega}_j - \omega_j'}\right]. \end{aligned} \tag{3.24}$$

This formula is, however, problematic since it requires a priori knowledge of the complex function and resonance points of the system. This information is not usually obtainable from optical spectra. Fortunately, the procedure of phase retrieval for meromorphic susceptibilities can be performed with another type of technique based on the maximum entropy model [77, 78], which is the topic of Chap. 5.

Before closing this chapter, we consider one example of a degenerate, meromorphic nonlinear third-order susceptibility, which has been introduced for the description of four-wave mixing in semiconductors. It can be written as follows [79]:

$$\chi^{(3)}(\omega, \omega, -\omega) = A\,|F(\omega)|^{-2}\,[F(\omega)]^{-2}, \tag{3.25}$$

where A is a constant. The real and imaginary parts of this nonlinear susceptibility are shown in Fig. 3.6, together with the corresponding curves computed using (3.24), but omitting the residues term. Clearly the contribution of the residues is significant and the Kramers–Kronig-type dispersion relations give only a rough approximation to the real and imaginary parts.

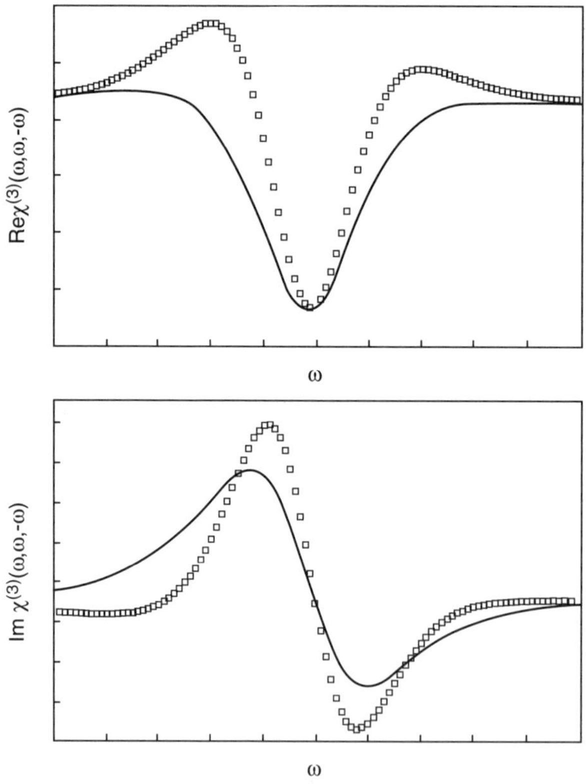

Fig. 3.6. Re $\left\{\chi^{(3)}(\omega, \omega, -\omega)\right\}$ and Im $\left\{\chi^{(3)}(\omega, \omega, -\omega)\right\}$ given by (3.25). The exact line shapes are shown by *solid lines* and the results of Kramers–Kronig analysis by *squares*

4. Conformal Mappings in Analysis of Optical Spectra

We have seen in the preceding chapters that optical spectra can be inverted by using appropriate dispersion relations. We also observed that there can be problems in their use since data extrapolations beyond the measured spectral range are usually needed. Therefore we may try to find alternative methods to obtain the desired optical constant or nonlinear quantity from the spectrum of a material. One method, whose advantage is its relative simplicity, is based on the application of conformal mapping to the interpretation of complex quantities in optical spectroscopy.

We start by investigating a simple conformal mapping and applying it to the description of optical constants and reflectance, exploiting their symmetry properties. The concept of conformal mapping is then applied to oblique-angle reflection spectroscopy, where the polarization of the light field has an important role. Finally, conformal mappings related to nonlinear susceptibilities are treated.

4.1 Conformal Mappings

Generally, a set of equations

$$U = U(x, y),$$

$$V = V(x, y) \tag{4.1}$$

defines a mapping which provides a correspondence between points belonging to the (x, y) and (U, V) planes. We are especially interested in holomorphic functions $f(z) = f(x + \mathrm{i}y) = U(x, y) + \mathrm{i}V(x, y)$ of the complex variable z. In that case, the Jacobian determinant of the mapping is

$$\begin{vmatrix} \dfrac{\partial U}{\partial x} & \dfrac{\partial U}{\partial y} \\ \dfrac{\partial V}{\partial x} & \dfrac{\partial V}{\partial y} \end{vmatrix} = \left| \frac{\mathrm{d}f}{\mathrm{d}z} \right|^2 . \tag{4.2}$$

The mapping is one to one if $\mathrm{d}f/\mathrm{d}z \neq 0$.

Conformal mapping means that two curves that intersect in the (x, y) plane are mapped into the (U, V) plane in such a manner that the angle between the curves is preserved. The nice feature of conformal mappings is that we can consider a function that is holomorphic in the original space as a holomorphic function in the new space.

There is one conformal mapping that has been found to be especially applicable in optical spectroscopy. It is the bilinear transformation

$$W = U + \mathrm{i}V = \frac{az + b}{cz + d}, \tag{4.3}$$

where a, b, c and d are constants and it is required that $ad - bc \neq 0$. The transformation given by (4.3) maps circles on the (x, y) plane into circles on the (U, V) plane. Furthermore, a half plane is mapped into a circle. Quite often we have to deal with holomorphic functions in a half plane, and then we may try to find a more appropriate description of them in a circle. Indeed, one such conformal mapping is the mapping of the upper half of the complex angular frequency plane into a unit disk by

$$W = \frac{\widehat{\omega} - \mathrm{i}}{\widehat{\omega} + \mathrm{i}}, \tag{4.4}$$

illustrated in Fig. 4.1. Now the real axis of the original plane is mapped into the boundary of the unit disk, whereas the upper half plane, which is the region of holomorphicity of, for example, the complex refractive index, is mapped inside the unit disk. Points located in the lower half plane are mapped to the exterior of the unit disk and, in particular, $\widehat{\omega} = -\mathrm{i}\,\mathrm{s}^{-1}$ is mapped to infinity. We wish to remark that when studying optical spectra we are actually working on the boundary of the unit disk, i.e. $|W| = 1$.

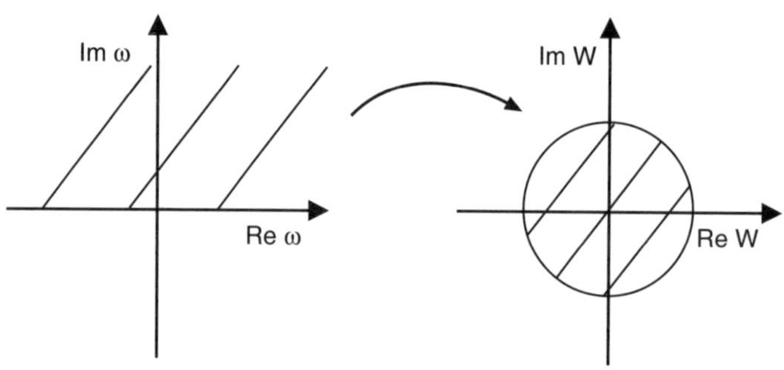

Fig. 4.1. Conformal mapping of the upper half plane into the unit disk

Another example of the bilinear transformation is the formula for the normal reflectance, which can be given as

$$r(\omega) = \frac{1 - N(\omega)}{1 + N(\omega)}$$

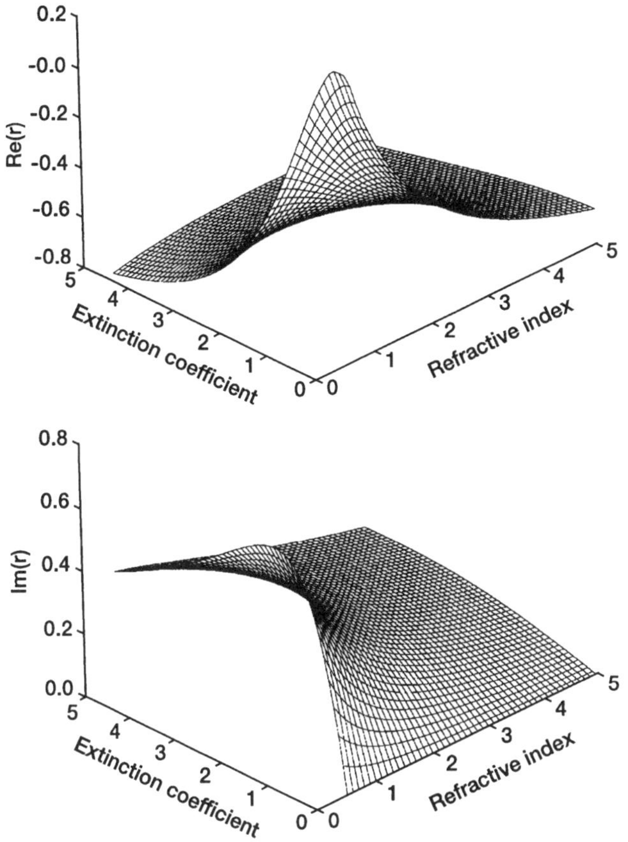

Fig. 4.2. Re $\{r(n,k)\}$ and Im $\{r(n,k)\}$ given by (4.5)

$$= \frac{1 - n^2(\omega) - k^2(\omega)}{[1 + n(\omega)]^2 + k^2(\omega)} - \mathrm{i}\frac{2k(\omega)}{[1 + n^2(\omega)]^2 + k^2(\omega)}. \tag{4.5}$$

Now if we investigate the real and imaginary parts of the electric-field reflectance in the (n, k) plane, we find results such as those in Fig. 4.2.

The concept of conformal mapping can be generalized to apply the description of the oblique-angle reflectances r_{s} and r_{p}, which will be discussed in Sect. 4.4.

4.2 Laurent Series Expansion of Complex Refractive Index in Unit Disk

Since the complex refractive index is a holomorphic function in the upper half plane, we can apply the bilinear transformation given by (4.4), whereafter the complex refractive index can be described as a holomorphic function in the unit disk. One of the basic results in complex analysis is related to the description of a holomorphic function by a Laurent series expansion. In the case of the complex refractive index of insulators, the Laurent series expansion is as follows:

$$N\left(-\mathrm{i}\frac{W+1}{W-1}\right) = \sum_{j=-\infty}^{\infty} c_j W^j, \tag{4.6}$$

where the coefficients c_j are complex numbers. However, in reality the imaginary angular frequency $\widehat{\omega} = \mathrm{i}\,\mathrm{s}^{-1}$ is not a singular point, and therefore the expansion can be simplified to the form

$$N\left(-\mathrm{i}\frac{W+1}{W-1}\right) = \sum_{j=0}^{\infty} c_j W^j. \tag{4.7}$$

Inspection of the symmetry properties of the original and transformed complex refractive indices is helpful since we can easily derive series expansions that assist us to find a method for optical data inversion. If we investigate what happens to W when we replace ω by $-\omega$, we find that

$$\frac{-\omega-\mathrm{i}}{-\omega+\mathrm{i}} = \frac{\omega+\mathrm{i}}{\omega-\mathrm{i}} = W^*. \tag{4.8}$$

Therefore the symmetry property of the complex refractive index is characterized by

$$\sum_{j=0}^{\infty} c_j W^{*j} = \left(\sum_{j=0}^{\infty} c_j W^j\right)^* . \tag{4.9}$$

The above result evidently implies that the coefficients c_j have to be real numbers. Since the physical angular frequencies are those that correspond to the boundary of the unit disk, we can substitute $W = \mathrm{e}^{\mathrm{i}\varphi}$ and obtain a more practical expression

$$N\left(-\mathrm{i}\frac{W+1}{W-1}\right) = \sum_{j=0}^{\infty} c_j \cos(j\varphi) + \mathrm{i}\sum_{j=0}^{\infty} c_j \sin(j\varphi). \tag{4.10}$$

After some mathematical manipulations we find also that

$$\omega = -\mathrm{i}\frac{\mathrm{e}^{\mathrm{i}\varphi}+1}{\mathrm{e}^{\mathrm{i}\varphi}-1} = -\mathrm{i}\frac{\mathrm{e}^{\mathrm{i}\varphi/2}+\mathrm{e}^{-\mathrm{i}\varphi/2}}{\mathrm{e}^{\mathrm{i}\varphi/2}-\mathrm{e}^{-\mathrm{i}\varphi/2}} = -\cot(\varphi/2). \tag{4.11}$$

Using this information we can define the complex refractive index by a new expression,

$$N\left[-\cot(\varphi/2)\right] - 1 = n(\varphi) - \mathrm{i}k(\varphi), \tag{4.12}$$

where the symmetry properties $k(-\omega) = -k(\omega)$ and $n(-\omega) = n(\omega)$ have been employed. Next we can separate the real and imaginary parts to obtain

$$n(\varphi) = c_0 + \sum_{j=1}^{\infty} c_j \cos(j\varphi),$$

$$k(\varphi) = \sum_{j=1}^{\infty} c_j \sin(j\varphi). \tag{4.13}$$

The last procedure yields conjugate Fourier series expansions, as shown by King [80]. Indeed, one special case of the Laurent series expansion is the complex form of the Fourier series expansion. In other words, when we limit the variable to the boundary of the unit disk, we have nothing but a Fourier series expansion expanded in the interval $(0, 2\pi)$. The important thing about the relations of (4.13) is that they both involve the same coefficients c_j. From this, we easily realize that if, for instance, the extinction coefficient is measured, we can fit the data by means of the second equation of (4.13). This means that we seek the appropriate coefficients c_j to obtain the best possible fit to the data. In practise we have to truncate the series expansions, i.e. set $c_j = 0$ when $j \geq m$. After this procedure we can construct the real refractive index using the calculated coefficients. However, one coefficient, c_0, still remains undetermined. This problem can be overcome by the following limiting process:

$$\lim_{\omega \to \infty} [n(\omega) - 1] = \lim_{\varphi \to 0} \{n[\cot(\varphi/2)] - 1\} = c_0 + \sum_{j=1}^{\infty} c_j = 0, \tag{4.14}$$

which implies that

$$c_0 = -\sum_{j=1}^{\infty} c_j. \tag{4.15}$$

The same procedure for optical data inversion, starting from conformal mapping and the Laurent series expansion, is valid also for metals, but the DC conductivity term has now to be taken into account. The information given by (4.15) is also important in that it provides information about the complex refractive index at the angular frequency $\widehat{\omega} = \mathrm{i\,s}^{-1}$,

$$N(\mathrm{i\,s}^{-1}) = c_0. \tag{4.16}$$

Another way to estimate the complex refractive index at the unit imaginary angular frequency is to apply either (2.38) or (2.41), as described in Sect. 2.5.

An illustration of the success of the above method is presented in Fig. 4.3, where the absorption bands and data-inverted real refractive index change of the mixed alkali halide crystals $\mathrm{KCl}_x\ \mathrm{Br}_{1-x}$, $0 \leq x \leq 1$, containing F color centers are shown [81]. The results of this data inversion match quite well with those obtained by the Kramers–Kronig relation. In practical calculations, the Fourier series coefficients are determined by fitting a finite number of data points by means of a fast Fourier transformation.

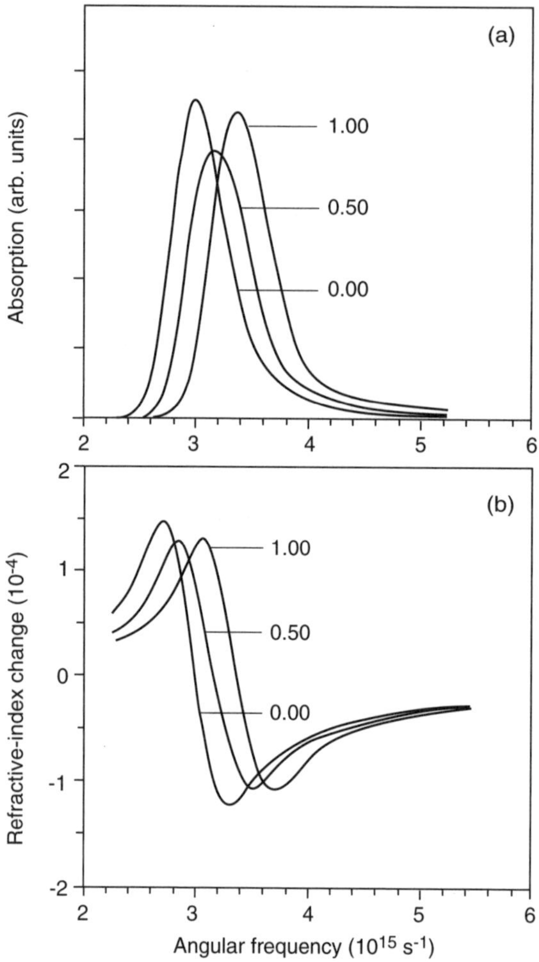

Fig. 4.3. Absorption and refractive index change for $KCl_x\ Br_{1-x}$ crystals containing F color centers. The values of x are indicated

4.3 Phase Retrieval Using Laurent Series Expansion of the Complex Derivative of Normal Reflectance

Phase retrieval with the aid of the conformal mapping of (4.4) and the Laurent series expansion is problematic for the reflectance at normal incidence. The reason is that since $\ln r(\omega)$ is not a bounded function in the interval $(0, \infty)$, the assumptions required in the use of Fourier series expansions are invalid. To overcome this problem King [82] derived a method that uses the derivative of the complex reflectivity, for which these assumption are valid. In addition, the derivative is holomorphic in the upper half plane (a holomorphic function belongs to the so-called C^{∞} class of functions, which means that the derivatives are also holomorphic functions) and is given by

$$-\mathrm{i}\frac{\mathrm{d}}{\mathrm{d}\widehat{\omega}}\ln r\left(\widehat{\omega}\right) = -\mathrm{i}\frac{\mathrm{d}}{\mathrm{d}\widehat{\omega}}\left[R^{1/2}\left(\widehat{\omega}\right)\mathrm{e}^{\mathrm{i}\varphi\left(\widehat{\omega}\right)}\right]$$

$$= \frac{\mathrm{d}\varphi\left(\widehat{\omega}\right)}{\mathrm{d}\widehat{\omega}} - \mathrm{i}\left[2R\left(\widehat{\omega}\right)\right]^{-1}\frac{\mathrm{d}R\left(\widehat{\omega}\right)}{\mathrm{d}\widehat{\omega}}. \tag{4.17}$$

Information about the last term on the right-hand side of this equation can be obtained by reflectance measurements. No information about the derivative of the phase angle is revealed by reflectance measurements. However, this derivative can be calculated as below.

Just as in the previous section, we exploit here the conformal mapping of (4.4) and write the Laurent series expansion

$$-\mathrm{i}\left[\frac{\mathrm{d}}{\mathrm{d}\widehat{\omega}}\ln r\left(\widehat{\omega}\right)\right]_{\widehat{\omega}=-\mathrm{i}(W+1)/(W-1)} = \sum_{j=0}^{\infty} c_j W^j, \tag{4.18}$$

and after the substitution $W = \mathrm{e}^{\mathrm{i}\Phi}$ we get

$$-\mathrm{i}\left[\frac{\mathrm{d}}{\mathrm{d}\widehat{\omega}}\ln r\left(\widehat{\omega}\right)\right]_{\widehat{\omega}=-\mathrm{i}(W+1)/(W-1)} = \sum_{j=0}^{\infty} c_j \cos\left(j\Phi\right) + \sum_{j=0}^{\infty} c_j \sin\left(j\Phi\right). \tag{4.19}$$

The procedure is quite similar to, but mathematically more complicated than, that in the previous section (here, we must take account of the symmetry properties of the reflectivity). As before, the coefficients c_j are real numbers and $c_0 = -\sum_{j=1}^{\infty} c_j$, which allows us to write the derivative of the phase angle for a real angular frequency:

$$\frac{\mathrm{d}\varphi\left(\omega\right)}{\mathrm{d}\omega} = \sum_{j=1}^{\infty} c_j\left[\cos\left(j\Phi\right) - 1\right]$$

$$= \sum_{j=1}^{\infty} c_j\left(2^{j-1}\cos^j\Phi + \sum_{l=1}^{[m/2]} b_{lm}\cos^{m-2l}\Phi\right), \tag{4.20}$$

where

$$b_{lm} = (-1)^l\left(\frac{m}{l}\right)2^{m-2l-1}\binom{m-l-1}{l-1} \tag{4.21}$$

represents a binomial coefficient. The upper summation limit $[m/2]$ in the series expansion inside the brackets on the right-hand side of (4.20) is equal to $\frac{1}{2}(m-1)$ if m is odd.

Next we solve for the real part using (4.4) under the assumption of a real angular frequency and find that

$$\cos\Phi = \frac{\omega^2 - 1}{\omega^2 + 1}. \tag{4.22}$$

After substitution of this value of $\cos\Phi$ into (4.20) we get

$$\frac{\mathrm{d}\varphi(\omega)}{\mathrm{d}\omega}=\sum_{j=1}^{\infty}c_j\left[-1+2^{m-1}\left(\frac{\omega^2-1}{\omega^2+1}\right)^m+\sum_{l=1}^{[m/2]}b_{lm}\left(\frac{\omega^2-1}{\omega^2+1}\right)^{m-2l}\right]. \tag{4.23}$$

Since our wish is to find the phase itself we integrate as follows:

$$\varphi(\omega)=\int_0^{\infty}\frac{\mathrm{d}\varphi(\omega)}{\mathrm{d}\omega}\mathrm{d}\omega$$

$$=\sum_{j=1}^{\infty}c_j\left[-\omega+2^{m-1}I_m(\omega)+\sum_{l=1}^{[m/2]}b_{lm}I_{m-2l}(\omega)\right], \tag{4.24}$$

where we have taken $\varphi(0)=0$ and $I_m(\omega)$ is defined by

$$I_m(\omega)=\int_0^{\omega}\left(\frac{\omega'^2-1}{\omega'^2+1}\right)^m\mathrm{d}\omega'. \tag{4.25}$$

The last integral can be expressed in closed form.

4.4 Conformal Mapping in Description of Oblique-Angle Reflectivity

It is obvious that the formula for the complex reflection coefficient r_s given by (1.27) can be treated as a bilinear transformation in the complex (a,b) plane. The coefficient r_p depends on r_s according to (1.32). These two quantities can be considered as holomorphic functions of $a+ib$ in the upper halves of complex planes described by $(\mathrm{Re}\{r_\mathrm{s}\},\ \mathrm{Im}\{r_\mathrm{s}\})$ and $(\mathrm{Re}\{r_\mathrm{p}\},\ \mathrm{Im}\{r_\mathrm{p}\})$. Such a discovery stimulated Azzam [16] to find various basic mathematical properties of the reflectivities for s- and p-polarized light at oblique incidence. As an example let us study the ratio

$$\frac{r_\mathrm{p}}{r_\mathrm{s}}=\frac{r_\mathrm{s}-\cos 2\theta}{1-r_\mathrm{s}\cos 2\theta}, \tag{4.26}$$

which is a bilinear transformation from the plane $(\mathrm{Re}\{r_\mathrm{s}\},\ \mathrm{Im}\{r_\mathrm{s}\})$ into the plane $(\mathrm{Re}\{r_\mathrm{p}\},\ \mathrm{Im}\{r_\mathrm{p}\})$. The mapping properties are illustrated in Fig. 4.4, where angularly spaced straight lines in the original space are mapped into curved lines in the $(\mathrm{Re}\{r_\mathrm{p}\},\ \mathrm{Im}\{r_\mathrm{p}\})$ plane. On the other hand, equispaced semicircles are mapped into spiral lines shown in Fig. 4.5.

Graphical nomograms for the determination of the phases of r_s and r_p can be constructed, which help in finding the optical constants of transparent and opaque materials. Graphical determination of the complex reflectivity is possible by means of the moduli of (4.26). Conformal mapping also provides a means to obtain the oblique-angle reflectance for s- and p-polarized light from the normal reflectance [83].

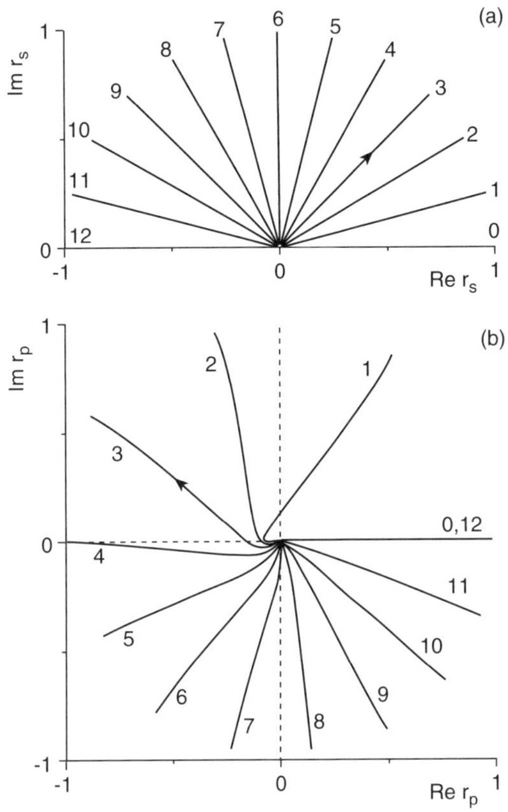

Fig. 4.4. Straight lines in the r_s plane **(a)**, mapped conformally into the r_p plane **(b)**. The angle of incidence is $\theta = 30°$ [16]

4.5 Conformal Mapping of Nonlinear Susceptibilities

The idea of using conformal mappings can be generalized to holomorphic nonlinear susceptibilities. The bilinear transformation of (4.4) is generalized to

$$W_j = \frac{\widehat{\omega}_j - \mathrm{i}}{\widehat{\omega}_j + \mathrm{i}}, \tag{4.27}$$

where $j = 1, 2, \ldots, n$ and the nonlinear susceptibility is assumed to be holomorphic in all appropriate half planes (either upper or lower). The real axes are mapped onto the boundaries of the individual unit disks $|W_j| = 1$, $j = 1, 2, \ldots, n$. As a result we have a function that is holomorphic in a polydisk. According to the theory of several complex variables this function can be expressed as a series expansion [64]:

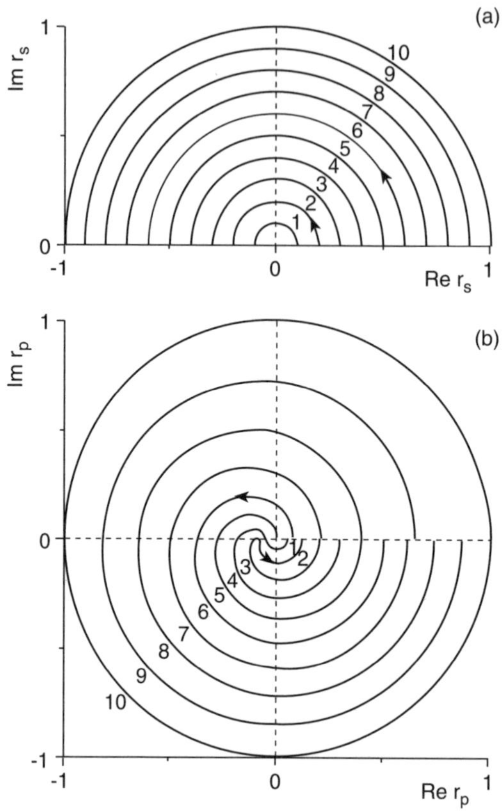

Fig. 4.5. Semicircles in the r_s plane **(a)**, mapped conformally into the r_p plane **(b)**. The angle of incidence is $\theta = 30°$ [16]

$$\chi^{(n)}(W_1, W_2, \ldots, W_n) = \sum_{\alpha=0}^{\infty}\sum_{\beta=0}^{\infty}\cdots\sum_{\nu=0}^{\infty} c_{\alpha\beta\cdots\nu} W_1^{\alpha} W_2^{\beta} \cdots W_n^{\nu}. \quad (4.28)$$

Using similar arguments about the symmetry properties to those used in Sects. 4.2 and 4.3, but now for the complex nonlinear susceptibility, we find that

$$\sum_{\alpha=0}^{\infty}\sum_{\beta=0}^{\infty}\cdots\sum_{\nu=0}^{\infty} c^{*}_{\alpha\beta\cdots\nu} W_1^{*\alpha} W_2^{*\beta} \cdots W_n^{*\nu}$$

$$= \sum_{\alpha=0}^{\infty}\sum_{\beta=0}^{\infty}\cdots\sum_{\nu=0}^{\infty} c_{\alpha\beta\cdots\nu} W_1^{*\alpha} W_2^{*\beta} \cdots W_n^{*\nu}, \quad (4.29)$$

which means that the coefficients of the series expansion are real numbers. Frequently, we deal with simpler nonlinear susceptibilities involving only pump and probe waves. Then the series expansion (4.28) reduces to a simpler expression. As an example let us study the properties of the third-order nonlinear Raman susceptibility, which has importance in coherent anti-Stokes Raman spectroscopy (CARS) and in Raman-induced Kerr effect spectroscopy (RIKES), which both involve four-wave mixing. The measured data usually contain information about the complex Raman susceptibility $\chi_{\mathrm{R}}^{(3)}$ and the background susceptibility $\chi_{\mathrm{NR}}^{(3)}$, which has usually a real value. The Lorentzian model can be used to describe the total third-order complex nonlinear susceptibility, in the vicinity of a single Raman mode, as follows:

$$\begin{aligned}\chi^{(3)} &= \chi_{\mathrm{NR}}^{(3)} + \chi_{\mathrm{R}}^{(3)} \\ &= \chi_{\mathrm{NR}}^{(3)} + \frac{A}{\omega_{\mathrm{R}} - (\omega_1 - \omega_2) + \mathrm{i}\Gamma},\end{aligned} \tag{4.30}$$

where the amplitude A is proportional to the Raman amplitude, ω_{R} is the Raman-mode angular frequency and ω_1 and ω_2 are the angular frequencies of the pump and probe lasers. If we set

$$\begin{aligned}W_0 &= \frac{\omega_{\mathrm{R}} - (\omega_1 - \omega_2)}{\Gamma}, \\ W &= \frac{W_0 - \mathrm{i}}{W_0 + \mathrm{i}},\end{aligned} \tag{4.31}$$

then we have a nonlinear susceptibility which can be expressed as a series expansion of one variable in a unit disk. In this particular case of (4.30), the series expansion has only two terms,

$$\chi^{(3)} = \chi_{\mathrm{NR}}^{(3)} - \frac{\mathrm{i}A}{2\Gamma}(1 - W). \tag{4.32}$$

In general, the nonlinear third-order Raman susceptibility, involving two external electric fields, can be given by the series expansion

$$\chi^{(3)}(W_1, W_2) = \sum_j \sum_l c_{jl} W_1^j W_2^l. \tag{4.33}$$

As usual, one variable is fixed (corresponding to the angular frequency of the pump laser). Then the series expansion of (4.33) can be considered as a series expansion $\sum_m c_m(W_2) W_1^m$ of one variable, where the coefficients $c_m(W_2)$ can be complex numbers. Then the nonlinear susceptibility is a holomorphic function of the variable W_1.

It is also possible to find an integral relation for the complex linear optical constants and the nonlinear susceptibility that is analogous to the Hilbert transform, but which involves integration along the boundary of the unit disk. Such a relation is obtained from the Poisson theory for a disk [29, 84]. Usually, however, the calculation of the corresponding Cauchy principal value of the integral is more problematic than that in the original angular frequency plane.

Most often, we gain information about only the intensity of a nonlinear optical signal. Then we have to deal with the modulus of the nonlinear susceptibility. The modulus of the nonlinear Raman susceptibility can be treated using the series expansion of (4.33). Before generalizing the dispersion relation to the modulus of the Raman susceptibility, now defined in a unit disk, we form the squared modulus of the Raman susceptibility of given by (4.32):

$$\left|\chi^{(3)}\right|^2 = \left(\chi_{\mathrm{NR}}^{(2)}\right)^2 + \frac{\mathrm{i}A\chi_{\mathrm{NR}}^{(3)}}{2\Gamma}\left(W - W^*\right) + \frac{A^2}{4\Gamma^2}\left(1 - W - W^* - WW^*\right). \tag{4.34}$$

We observe that the holomorphicity related to the squared modulus of the nonlinear susceptibility may have some complications due to the complex conjugates of the variable. According to the theory of quasiconformal mappings [85], however, we can formulate a generalized Cauchy formula, which is given by [84]

$$\left|\chi^{(3)}\left(W_1', W_2\right)\right|^2 = \frac{1}{\mathrm{i}\pi}\mathrm{P}\int\limits_{|W_1|=1} \frac{\left|\chi^{(3)}\left(W_1, W_2\right)\right|^2}{W_1 - W_1'}\mathrm{d}W_1$$

$$-\frac{1}{\pi}\iint_{|W_1|\leq 1} \frac{\mathrm{d}/\mathrm{d}W_1^*\left[\left|\chi^{(3)}\left(W_1, W_2\right)\right|^2\right]}{W_1 - W_1'}\mathrm{d}\left(\mathrm{Re}\{W_1\}\right)\mathrm{d}\left(\mathrm{Im}\{W_1\}\right), \tag{4.35}$$

where P denotes the Cauchy principal value just as before and there is a pole on the boundary of the unit circle, as shown in Fig. 4.6. However, now we must include, in addition to the familiar dispersion integral, an extra term on the right-hand side of the equation. Indeed, a surface integral appears, and the complex derivative in the integrand is defined by

$$\frac{\mathrm{d}}{\mathrm{d}W_1^*} = \frac{1}{2}\left[\frac{\mathrm{d}}{\mathrm{d}\left(\mathrm{Re}\left\{W_1\right\}\right)} + \mathrm{i}\frac{\mathrm{d}}{\mathrm{d}\left(\mathrm{Im}\left\{W_1\right\}\right)}\right]. \tag{4.36}$$

The calculation of the surface integral is outlined in Appendix F.

In the case of the squared modulus of the Raman susceptibility given in (4.34), the contour integral of (4.35) gives the value $\left(\chi_{\mathrm{NR}}^{(3)}\right)^2 + R^2/(2\Gamma)$, whereas the surface integral yields $R^2/(2\Gamma)$, for a pole at resonance. Evidently the surface integral cannot be neglected. Unfortunately, it is not possible by measurement to get information about either the squared modulus inside the unit disk or the complex derivative, information about which is required in (4.35). Therefore we have to seek more practical ways to carry out the phase retrieval procedure. In addition, the phase retrieval is problematic because the measured spectra cover only a finite range of angular frequencies. Therefore another type of method, rather than that based on the use of (4.33) and (4.35), is more convenient. In the next section we describe such a method in detail.

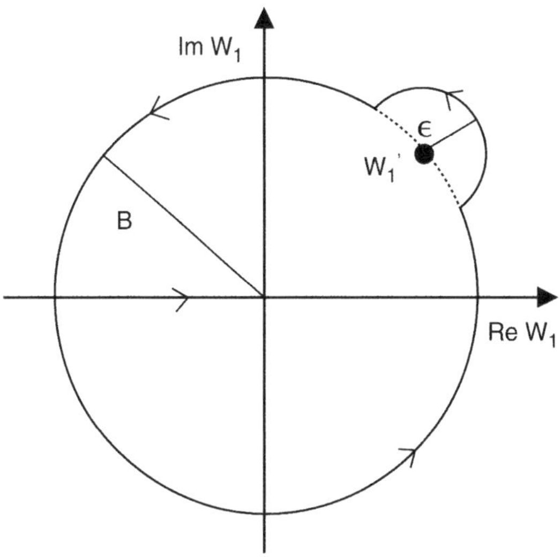

Fig. 4.6. Contour used for integral in (4.35); a complex pole W'_1 is located on the contour

4.6 Conformal Mapping in Analysis of the Modulus of Nonlinear Susceptibility

We consider here only the nonlinear Raman susceptibility mentioned in the previous section. In the case of CARS and RIKES, the intensity of the spectrum is proportional to the squared modulus of the nonlinear susceptibility or its effective value. Then the role of the background susceptibility is not negligible as concerns the line shape of the spectrum. The squared modulus can be written as

$$\left|\chi^{(3)}\right|^2 = \left(\chi^{(3)}_{\mathrm{NR}}\right)^2 + 2\chi^{(3)}_{\mathrm{NR}}\,\mathrm{Re}\left\{\chi^{(3)}_{\mathrm{R}}\right\} + \left|\chi^{(3)}_{\mathrm{R}}\right|^2 . \tag{4.37}$$

We have in practise a finite angular frequency range $(-W'_J, W'_J)$, and choose a variable $W' = \omega_1 - \omega_2 - \omega_{\mathrm{R}}$ for the sake of convenience. Previously we mapped the whole upper half plane into the unit disk. Here we employ a mapping that maps a semi-infinite strip of the upper half plane into a unit disk, which is supplied by

$$W = \mathrm{e}^{\mathrm{i}\pi W'/W_J} . \tag{4.38}$$

The finite interval $(-W'_J, W'_J)$ of the real axis is mapped into the boundary of the unit disk. The upper half plane can be imagined to be divided into semi-infinite strips of equal width. Such a division results in a periodicity of the optical spectrum because of the mapping of (4.38). However, in reality there is usually no specific periodicity in optical spectra. In practical calculations we limit our consideration to a single semi-infinite strip. In addition, the series expansion is truncated, i.e. we approximate the nonlinear susceptibility by the polynomial

$$\chi_{\mathrm{R}}^{(3)}(\varphi) = \sum_{j=0}^{J} c_j \mathrm{e}^{\mathrm{i}j\varphi}. \tag{4.39}$$

The restriction to a polynomial arises from the fact that when we form the squared modulus of the nonlinear susceptibility, we must multiply two series. Such a multiplication is possible for series with an infinite sum of terms only if the series expansion is absolutely convergent on the boundary of the unit disk. We do not know of such a convergence in the general case of any measured line shape. To avoid this problem, we can approximate the nonlinear susceptibility at the boundary of the unit disk by a finite series expansion, i.e. by a polynomial. The number of terms appearing in (4.39) can always be increased in order to obtain a good numerical approximation.

Let us first assume, for the sake of generality, that the coefficients in the series expansion of (4.39) are complex numbers. The squared modulus of $\chi_{\mathrm{R}}^{(3)}$ can be approximated as follows:

$$\begin{aligned}\left|\chi_{\mathrm{R}}^{(3)}(\varphi)\right|^2 &= \left(\sum_{j=0}^{J} c_j \mathrm{e}^{\mathrm{i}j\varphi}\right)\left(\sum_{j=0}^{J} c_j^* \mathrm{e}^{-\mathrm{i}j\varphi}\right) \\ &= A_0' + 2\sum_{j=1}^{J}\left[A_j' \cos(j\varphi) + B_j' \sin(j\varphi)\right],\end{aligned} \tag{4.40}$$

where

$$\begin{aligned} A_j' &= \sum_{l=0}^{J-j}(a_l a_{l+j} + b_l b_{l+j}),\ j = 0, 1, \ldots, J, \\ B_j' &= \sum_{l=0}^{J-j}(a_{l+j} b_l - a_l b_{l+j}),\ j = 1, 2, \ldots, J-1. \end{aligned} \tag{4.41}$$

A detailed derivation of (4.40) is given in Appendix G. If we assume that $\mathrm{Re}\left\{\chi_{\mathrm{R}}^{(3)}\right\}$ is an odd function of W', then the coefficients a_j are identically zero, because the Fourier series expansion of $\mathrm{Re}\left\{\chi_{\mathrm{R}}^{(3)}\right\}$ involves only sine terms. Then we can write

$$\chi_{\mathrm{R}}^{(3)}(\varphi) = \sum_{j=0}^{J} \mathrm{i}b_j \mathrm{e}^{\mathrm{i}j\varphi},$$

$$\mathrm{Re}\left\{\chi_{\mathrm{R}}^{(3)}(\varphi)\right\} = -\sum_{j=1}^{J} b_j \sin(j\varphi),$$

$$\left|\chi_{\mathrm{R}}^{(3)}(\varphi)\right|^{2}=A_{0}+2\sum_{j=1}^{J}A_{j}\cos\left(j\varphi\right), \tag{4.42}$$

where

$$A_{j}=\sum_{l=0}^{J-j}b_{l}b_{l+j}, j=0,1,\ldots,N.$$

By substituting the formulae of (4.42) into (4.37), we obtain

$$\left|\chi^{(3)}(\varphi_{m})\right|^{2}=\alpha_{0}+2\sum_{j=1}^{J-1}\left[\alpha_{j}\cos\left(j\varphi_{m}\right)+\beta_{j}\sin\left(j\varphi_{m}\right)\right]+\alpha_{J}\cos\left(\pi m\right), \tag{4.43}$$

where

$$\alpha_{0}=\left(\chi_{\mathrm{NR}}^{(3)}\right)^{2}+A_{0}, \alpha_{j}=2A_{J},$$

$$\alpha_{j}=A_{j}, j=1,2,\ldots,J-1,$$

$$\beta_{j}=-\chi_{\mathrm{NR}}^{(3)}b_{j}, j=1,2,\ldots,J-1. \tag{4.44}$$

Equation (4.43) represents a finite Fourier series in the set $\varphi_m=\pi m/J, m=-J+1,\ldots,-1,0,+1,\ldots,J$. The coefficients α_j and β_j can be given by discrete Fourier transformation formulae by using the measured data as follows:

$$\alpha_{j}=\frac{1}{2J}\sum_{m=-J+1}^{J}\left|\chi^{(3)}(\varphi_{m})\right|^{2}\cos\left(j\varphi_{m}\right), j=0,1,\ldots,J,$$

$$\beta_{j}=\frac{1}{2J}\sum_{m=-J+1}^{J}\left|\chi^{(3)}(\varphi_{m})\right|^{2}\sin\left(j\varphi_{m}\right), j=1,2,\ldots,J-1. \tag{4.45}$$

From (4.42) and (4.44), and using the coefficients of (4.45), we get the approximations

$$\left|\chi_{\mathrm{R}}^{(3)}(\varphi_{m})\right|^{2}+\left(\chi_{\mathrm{NR}}^{(3)}\right)^{2}=\alpha_{0}+2\sum_{j=1}^{J-1}\left[\alpha_{j}\cos\left(j\varphi_{m}\right)\right]+\alpha_{J}\cos\left(\pi m\right),$$

$$\chi_{\mathrm{NR}}^{(3)}\operatorname{Re}\left\{\chi_{\mathrm{R}}^{(3)}(\varphi_{m})\right\}=\sum_{j=1}^{J-1}\beta_{j}\sin\left(j\varphi_{m}\right). \tag{4.46}$$

Therefore we can calculate quantities that are directly proportional to the real part and the squared modulus of the nonlinear Raman susceptibility.

Regarding the imaginary part, we may use the formulae in (4.46) and the equality $\left|\chi_{\mathrm{R}}^{(3)}\right|^{2}=\left[\operatorname{Re}\left\{\chi_{\mathrm{R}}^{(3)}\right\}\right]^{2}+\left[\operatorname{Im}\left\{\chi_{\mathrm{R}}^{(3)}\right\}\right]^{2}$ or the Fourier coefficients of

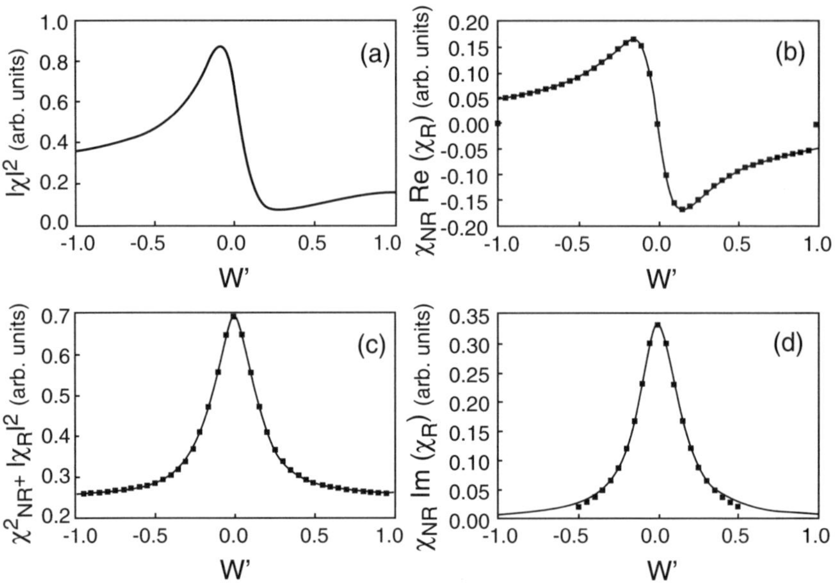

Fig. 4.7. **(a)** Squared modulus $\left|\chi^{(3)}\right|^2$ of the total third-order susceptibility given by (4.30) using parameter values $A = 0.1$, $\chi_{\mathrm{NR}} = 0.5$ and $\Gamma = 0.15$. **(b)**, **(c)** and **(d)**: curves of $\chi^{(3)}_{\mathrm{NR}} \operatorname{Re}\left\{\chi^{(3)}_{\mathrm{R}}\right\}$, $\left(\chi^{(3)}_{\mathrm{NR}}\right)^2 + \left|\chi^{(3)}_{\mathrm{R}}\right|^2$ and $\chi^{(3)}_{\mathrm{NR}} \operatorname{Im}\left\{\chi^{(3)}_{\mathrm{R}}\right\}$, respectively [86]

(4.45) to calculate quantities proportional to the imaginary part. An estimate can be given as follows:

$$\chi^{(3)}_{\mathrm{NR}} \operatorname{Im}\left\{\chi^{(3)}_{\mathrm{R}}(\varphi_m)\right\} = -\frac{\beta_0}{2} - \sum_{j=1}^{J-1} \beta_j \cos(j\varphi_m) - \frac{\beta_J}{2}(-1)^m . \qquad (4.47)$$

The coefficients β_0 and β_J are unknown. We may make the approximation $\beta_J \approx 0$, because the susceptibility is well approximated by the series expansion if J is large enough. On the other hand, for the resonance we have

$$b_0 \simeq \operatorname{Im}\left\{\chi^{(3)}_{\mathrm{R}}(0)\right\} - \sum_{j=1}^{J-1} b_j , \qquad (4.48)$$

where the background susceptibility is assumed to be known. If the background susceptibility is not known, it can be estimated by extrapolating from the curve given by the first formula of (4.46), taking into account the fact that

$$\lim_{W' \to \infty} \left|\chi^{(3)}_{\mathrm{R}}(W')\right| = 0. \qquad (4.49)$$

The results of the procedure described above are illustrated in Fig. 4.7 for a case of extracting the real and imaginary parts of a nonlinear Raman susceptibility of the form (4.30) when only the squared modulus of the total susceptibility is known [86].

The numerical method described above is powerful in the case of relatively simple line shapes in nonlinear optical spectra. In the next chapter we will introduce another method, based on the fitting of spectra by the maximum entropy method, which is powerful in phase retrieval problems in linear and nonlinear optics. In addition it is applicable to data analysis in the case of meromorphic nonlinear susceptibilities.

5. Maximum Entropy Method

Practical experiments in optics usually provide only partial measurements of the electric field $E(t)$. For example, in optical spectroscopy only an intensity spectrum, $|E(\omega)|^2 \propto |f(\omega)|^2$, is measured, while the entire complex response function, $f(\omega) = |f(\omega)| \exp[i\varphi(\omega)]$, is needed in order to obtain the desired material properties. This leads to the question of phase retrieval: with some extra obtainable information, is it possible to compute the phase function $\varphi(\omega)$ from a measurement of the amplitude $|f(\omega)|$? Recently, significant progress has been made in solving this problem in spectroscopic applications. This progress has been due mainly to the idea of applying a certain type of maximum entropy model to phase retrieval. This approach was first shown to be applicable to the analysis of coherent anti-Stokes Raman (CARS) spectra [87] and to reflection spectroscopy [88, 89]. Later, an improved version of the procedure, applicable to any case of spectrum analysis where $f(\omega)$ is to be computed from $|f(\omega)|$, was given by Vartiainen, Asakura and Peiponen [90]. In particular, this method was shown to be valid even in the situation in which the alternative Kramers–Kronig method is not, i.e when $f(\widehat{\omega})$ is a meromorphic complex function, such as the degenerate third-order nonlinear susceptibility $\chi^{(3)}(\widehat{\omega}, \widehat{\omega}, -\widehat{\omega})$ [77].

In this chapter we introduce the method of phase retrieval in optical spectroscopy using the maximum entropy model.

5.1 Maximum Entropy Model

It has been conjectured in information theory [91–93] that entropy defines a measure on the space of probability distributions, such that those of high entropy are favored over others. Accordingly, the maximum entropy (ME) principle states that any inferences made from incomplete information should be based on the probability distribution that has the maximum entropy permitted by the available data. Most notably, this principle has been utilized with iterative ME algorithms in various image restoration problems arising in astronomy [94, 95], X-ray crystallography [96], medical tomography [97] and electron holography [98].

There exists also another type of maximum entropy method, which was originally developed by Burg [99]. His method is an algorithm for computing the power spectrum $S(\omega)$ when the measured data consist of a sampled signal

$$x_m \equiv x(t_m), \quad 0 \leq m \leq M, \tag{5.1}$$

whose autocorrelation values are given by

$$C(m) = \varepsilon[x_n x_{n+m}], \tag{5.2}$$

where ε denotes the expectation operator. The autocorrelation function $C(m)$ and the power spectrum $S(\omega)$ form a Fourier transform pair,

$$S(\omega) = \Delta t \sum_{m=-\infty}^{\infty} C(m) \exp(-im\omega\Delta t), \tag{5.3}$$

$$C(m) = \frac{1}{2\pi} \int_0^{2\pi/\Delta t} S(\omega) \exp(im\omega\Delta t)\, \mathrm{d}\omega, \tag{5.4}$$

where Δt is the time between equispaced samples. In the following it is more convenient to use a normalized frequency $\nu = (2\pi)^{-1}\omega\Delta t$ (i.e. ν is normalized within the interval $0 \leq \nu \leq 1$). Equations (5.3) and (5.4) can now be written as

$$S(\nu) = \sum_{m=-\infty}^{\infty} C(m) \exp(-\mathrm{i}2\pi m\nu), \tag{5.5}$$

$$C(m) = \int_0^{1/\Delta t} S(\nu) \exp(\mathrm{i}2\pi m\nu)\, \mathrm{d}\nu, \tag{5.6}$$

where $S(\nu) = S(\omega)/\Delta t$.

The conventional method of obtaining a power spectrum estimate from known autocorrelation values $C(m)$, $|m| \leq M$, is to assume that $C(m) = 0$ for $|m| > M$ and to take a Fourier transform of $w(m)C(m)$, where $w(m)$ is a window function. The windowing is used to reduce the errors due to the truncation of the autocorrelation function. A shortcoming of this kind of estimation is the fact that there is no knowledge about the window function that would give the best result. In Burg's maximum entropy method, the idea is to choose the spectrum that corresponds to the most random or the most unpredictable time series whose autocorrelation function agrees with the set of known values. This leads to an extrapolation of the autocorrelation function by maximizing the entropy (a measure of the average information content) of the process $x(t)$; hence the name of the method. As a consequence, Burg's procedure leads to a certain maximum entropy model for $S(\omega)$. On the one hand, this model is the final result of Burg's approach, and on the other hand, it is the starting point in the phase retrieval procedure described in Sect. 5.4. We emphasize that this model for $S(\omega)$ is the only feature of the phase retrieval method that depends on Burg's maximum entropy method.

For any stationary time series $\{x_n\}$ it can be shown that the entropy rate of the time series is [100]

$$h \propto \int_0^1 \log S(\nu)\, \mathrm{d}\nu, \tag{5.7}$$

where it is assumed that the time series is limited to the angular frequency band $0 \le \omega \le \omega_\mathrm{t}$ and $\omega_\mathrm{t} = 2\pi/\Delta t$.

Suppose that we know the first $2M + 1$ values of the autocorrelation function, namely

$$C(m) = \int_0^1 S(\nu) \exp(\mathrm{i}2\pi m\nu)\, \mathrm{d}\nu, \quad |m| \le M. \tag{5.8}$$

Then the most reasonable choice for the unknown values of $C(m)$, $|m| > M$, is the one that adds no information or entropy to the process. Therefore, the required maximum-entropy power spectrum is the function $S(\nu)$ that maximizes (5.7) under the constraint (5.8). That is,

$$\frac{\partial}{\partial C(m)} \int_0^1 \log S(\nu)\, \mathrm{d}\nu = 0, \quad |m| > M. \tag{5.9}$$

From (5.5) we see that

$$\frac{\partial S(\nu)}{\partial C(m)} = \exp(-\mathrm{i}2\pi m\nu). \tag{5.10}$$

Furthermore, it is also true that

$$\frac{\partial \log S(\nu)}{\partial C(m)} = \frac{1}{S(\nu)} \frac{\partial S(\nu)}{\partial C(m)} = [S(\nu)]^{-1} \exp(-\mathrm{i}2\pi m\nu). \tag{5.11}$$

Making use of (5.11), we see that (5.9) reduces to

$$\int_0^1 [S(\nu)]^{-1} \exp(-\mathrm{i}2\pi m\nu)\, \mathrm{d}\nu = 0, \quad |m| > M. \tag{5.12}$$

This equation specifies the form of the inverse power spectrum $[S(\nu)]^{-1}$ of the maximum-entropy process. The inverse power spectrum can be written as

$$[S(\nu)]^{-1} = \sum_{k=-\infty}^{\infty} a_k \exp(\mathrm{i}2\pi k\nu) = \sum_{k=-M}^{M} a_k \exp(\mathrm{i}2\pi k\nu), \tag{5.13}$$

where

$$a_k = \int_0^1 [S(\nu)]^{-1} \exp(-\mathrm{i}2\pi k\nu)\,\mathrm{d}\nu, \quad |k| \le M. \tag{5.14}$$

If we replace k by $-k$ in (5.13), we get also

$$[S(\nu)]^{-1} = \sum_{k=-M}^{M} a_k \exp(-\mathrm{i}2\pi k\nu), \tag{5.15}$$

where

$$a_k = \int_0^1 [S(\nu)]^{-1} \exp(\mathrm{i}2\pi k\nu)\,\mathrm{d}\nu, \quad |k| \le M. \tag{5.16}$$

Now, the maximum-entropy power spectrum estimate has the following form:

$$S(\nu) = \frac{1}{\sum_{k=-M}^{M} a_k \exp(-\mathrm{i}2\pi k\nu)}. \tag{5.17}$$

Next we need to determine values for the set of coefficients $\{a_k\}$ such that the power spectrum estimate $S(\nu)$ is consistent with the known values of the autocorrelation function, that is,

$$C(m) = \int_0^1 S(\nu) \exp(\mathrm{i}2\pi m\nu)\,\mathrm{d}\nu, \quad 0 \le m \le M. \tag{5.18}$$

Substituting (5.17) into (5.18), we get

$$C(m) = \int_0^1 \frac{\exp(\mathrm{i}2\pi m\nu)}{\sum_{k=-M}^{M} a_k \exp(-\mathrm{i}2\pi k\nu)}\mathrm{d}\nu, \quad 0 \le m \le M. \tag{5.19}$$

Now, by using the notation $z = \exp(\mathrm{i}2\pi\nu)$, we have $\mathrm{d}\nu = \mathrm{d}z/(\mathrm{i}2\pi z)$. Equation (5.19) may be rewritten in the form

$$C(m) = \frac{1}{2\pi\mathrm{i}} \oint_{|z|=1} \frac{z^{m-1}}{\sum_{k=-M}^{M} a_k z^{-k}}\mathrm{d}z, \quad 0 \le m \le M. \tag{5.20}$$

The series expansion in (5.13) can be written as the product of two polynomials (see Appendix H):

$$\sum_{k=-M}^{M} a_k z^{-k} = A_M(z)\, A_M^*(1/z^*), \tag{5.21}$$

where

$$
\begin{aligned}
&A_M(z) = \sum_{k=0}^{M} \alpha_k z^{-k}, \\
&A_M^*(1/z^*) = \sum_{k=0}^{M} \alpha_k^* z^k, \\
&a_k = \sum_{j=0}^{M-k} \alpha_j^* \alpha_{j+k}, \quad 0 \le k \le M, \\
&a_{-k} = a_k^*.
\end{aligned} \tag{5.22}
$$

Then (5.20) can be written as

$$
C(m) = \frac{1}{2\pi \mathrm{i}} \oint\limits_{|z|=1} \frac{z^{m-1}}{A_M(z) A_M^*(1/z^*)} \mathrm{d}z. \tag{5.23}
$$

Using (5.21), we get the summation

$$
\begin{aligned}
\sum_{j=0}^{M} \alpha_j C(m-j) &= \frac{1}{2\pi \mathrm{i}} \oint\limits_{|z|=1} \frac{z^{m-1} \sum\limits_{j=0}^{M} \alpha_j z^{-j}}{A_M(z) A_M^*(1/z^*)} \mathrm{d}z \\
&= \frac{1}{2\pi \mathrm{i}} \oint\limits_{|z|=1} \frac{z^{m-1} A_M(z)}{A_M(z) A_M^*(1/z^*)} \mathrm{d}z \\
&= \frac{1}{2\pi \mathrm{i}} \oint\limits_{|z|=1} \frac{z^{m-1}}{A_M^*(1/z^*)} \mathrm{d}z, \quad 0 \le m \le M.
\end{aligned} \tag{5.24}
$$

The integrand in (5.24) is holomorphic on the boundary of and inside the unit disk for $m \ge 1$, because the polynomial $A_M^*(1/z^*)$ is chosen to have its zeros all located outside the unit disk. So, according to Cauchy's residue theorem, the contour integral in (5.24) is zero for $m \ge 1$. However, for $m = 0$ the integrand has a simple pole at $z = 0$ with a residue equal to $1/\alpha_0^*$. Thus, we get

$$
\sum_{j=0}^{M} \alpha_j C(m-j) = \begin{cases} 1/\alpha_0^* & , \ m = 0 \\ 0 & , \ m = 1, 2, ..., M. \end{cases} \tag{5.25}
$$

By dividing both sides of (5.25) by α_0, we get the following set of linear equations:

$$
\begin{pmatrix} C(0) & C(-1) & \cdots & C(-M) \\ C(1) & C(0) & \cdots & C(1-M) \\ \vdots & \vdots & \ddots & \vdots \\ C(M) & C(M-1) & \cdots & C(0) \end{pmatrix} \begin{pmatrix} 1 \\ c_1 \\ \vdots \\ c_M \end{pmatrix} = \begin{pmatrix} b_0 \\ 0 \\ \vdots \\ 0 \end{pmatrix}, \tag{5.26}
$$

where $b_0 = |\alpha_0|^{-2}$ and $c_k = \alpha_k/\alpha_0$, $k = 1, 2, ..., M$. Substituting the coefficients c_k into (5.22), we obtain

$$A_M(z) = \alpha_0 \left(1 + \sum_{k=1}^{M} c_k z^{-k}\right),$$

$$A_M^*(1/z^*) = \alpha_0^* \left(1 + \sum_{k=1}^{M} c_k^* z^k\right), \tag{5.27}$$

and furthermore it follows from (5.21) that

$$\sum_{k=-M}^{M} a_k z^{-k} = b_0^{-1} \left|1 + \sum_{k=1}^{M} c_k z^{-k}\right|^2. \tag{5.28}$$

Finally, using (5.28) and (5.17) with $z = \exp(\mathrm{i}2\pi\nu)$, we get the maximum-entropy power spectrum estimate in the form

$$S(\nu) = \frac{b_0}{\left|1 + \sum_{k=1}^{M} c_k \exp(-\mathrm{i}2\pi k\nu)\right|^2}, \tag{5.29}$$

where the unknown coefficients c_k and b_0 can now be computed from the known autocorrelations $C(m)$, $|m| \le M$, by using (5.26).

Van den Bos [101] has shown that the maximum entropy model (MEM), as defined in (5.26) and (5.29), is equivalent to the autoregressive (AR) model of a stationary stochastic process $x(t)$. A time process is called autoregressive if an observation x_m can be given by a linear combination of the M preceding observations:

$$x_m = -\sum_{k=l}^{M} c_k x_{m-k} + e_m, \tag{5.30}$$

where e_m is the corresponding error at m (with $\varepsilon[e(n)e(m)] = 0$ for $n \ne m$), and the c_ks are the same coefficients as in (5.26) and (5.29).

Here we are interested in using the model of (5.29) for phase retrieval. Therefore, in addition to the model for $S(\nu) = |f(\nu)|^2$, we must have an ME model for $f(\nu) = |f(\nu)| \exp[\mathrm{i}\varphi(\nu)]$. We can derive this by taking a z-transform of both sides of (5.30) and using the convolution theorem to obtain

$$X(z) = -\left(c_1 z^{-1} + c_2 z^{-2} + \cdots + c_M z^{-M}\right) X(z) + E(z), \tag{5.31}$$

where $X(z)$ and $E(z)$ are the z-transforms of the sequences $\{x_m\}$ and $\{e_m\}$, respectively. Thus, we get the following expression for $X(z)$:

$$X(z) = \frac{E(z)}{1 + \sum_{k=l}^{M} c_k z^{-k}}. \tag{5.32}$$

Finally, if we replace the signal $x(t)$ with the time response function $g(t)$ and set $z = \exp(\mathrm{i}2\pi\nu)$, the z-transforms can be recognized as the discrete-time Fourier transforms. Hence, we can write the maximum-entropy estimate for the complex spectral response $f(\nu) = \mathcal{F}\{g(t)\}$ as a function of the normalized frequency ν as

$$f(\nu) = \frac{E(\nu)}{1 + \sum_{k=l}^{M} c_k \exp(-\mathrm{i}2\pi k\nu)}. \tag{5.33}$$

5.2 Phase Retrieval Procedure

Suppose that a power spectrum $S(\omega) = |f(\omega)|^2$ is measured within the frequency range $\omega_1 \leq \omega \leq \omega_2$. Now, by defining the normalized frequency as

$$\nu = (\omega - \omega_1) / (\omega_2 - \omega_1), \tag{5.34}$$

$S(\omega)$ can be fitted by (5.6), (5.26) and (5.29). However, instead of using the continuous Fourier transform of (5.6), the autocorrelations $C(m)$ are in practise computed by using the discrete Fourier transform:

$$C(m) = N^{-1} \sum_{n=0}^{N-1} S_n \exp(\mathrm{i}2\pi mn/N), \tag{5.35}$$

where N is the number of samples of $S(\omega_n) \equiv S_n$. In this case there exists an upper limit to the number of values of $C(m)$ and, therefore, to the MEM parameter M in (5.26) and (5.29), i.e. $M \leq N/2$. Unlike Burg's maximum entropy method, there is here no ambiguity in selecting the value of M. Namely, if the object is to make the MEM estimate $\widehat{S}(\omega)$ as close as possible to the original spectrum, then using the maximum value, $M = M_{\max}$, gives the best result. However, if the original spectrum $S(\omega)$ is noisy, it is possible to reduce the noise in $\widehat{S}(\omega)$ by computing it with a lower value of M. This is illustrated in Fig. 5.1 [90]. Three simulated spectra, with the same spectral features but different amounts of noise, are shown in Fig. 5.1a. The number of samples was selected to be $N = 1001$ and, consequently, $M_{\max} = 500$. Figure 5.1b shows how the value of M affects the root-mean-square (RMS) errors in the MEM estimates of the spectra in Fig. 5.1a. The RMS error is defined here as

$$E_{\mathrm{RMS}}^2 = \left(\int \left| S_M^{(i)}(\nu) - S_0(\nu) \right|^2 \mathrm{d}\nu \right) \left(\int |S_0(\nu)|^2 \mathrm{d}\nu \right)^{-1}, \tag{5.36}$$

where $S_M^{(i)}(\nu)$, $i = A, B, C$, is an MEM estimate computed from spectrum i in Fig. 5.1a, and $S_0(\nu)$ is the original noiseless spectrum (curve A in Fig. 5.1a). In the noise-free case of curve A, the minimum error $\left(E_{\mathrm{RMS}} \approx 10^{-6}\right)$ is obtained with the maximum value $M = 500$. The error increases with a decrease of M, although the increase is only nominal over a wide range of M. At low values of M less than 70, however, the error increases very strongly. The latter observation is also true for the noisy spectra of curves B and C. For noisy spectra, the smallest error (compared with the noiseless spectrum) is obtained using an optimum value $M = M_{\text{opt}}$ that is a compromise between high values of M which also reconstruct the noise (see Fig. 5.1c) and low values which do not result in much noise but result in distortions instead, because a higher-order MEM is required. Therefore, in practise, when M_{opt} cannot be determined, we should clearly use a high value of M to ensure it is greater than M_{opt}, rather than a low value less than M_{opt}, to avoid distortions at the expense of reconstructing some noise.

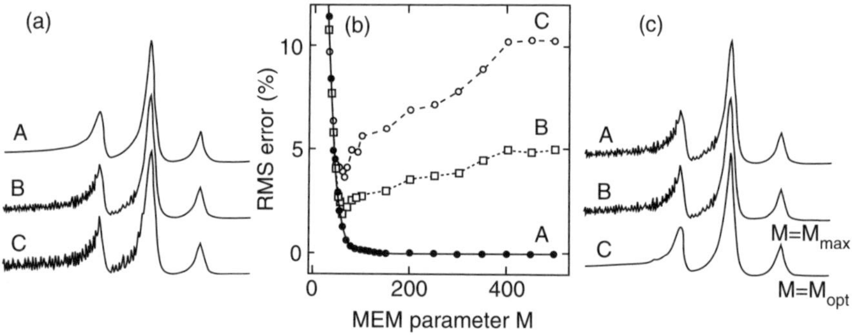

Fig. 5.1. **(a)** Simulated power spectra $S(\omega)$: original noiseless spectrum (curve A) and with noise-to-signal (N/S) ratios (defined as RMS values) of 0 (curve A), 5% (curve B) and 10% (curve C). **(b)** RMS errors versus value of M in MEM estimates of the spectra in (a). **(c)** Power spectrum $S(\omega)$ with $N/S = 5\%$ (curve A) and its MEM fits $\widehat{S}_M(\omega)$ with $M = 500$ (curve B) and $M = 60$ (curve C) [90]

The next step in the procedure is to find a way to compute $\widehat{f}(\nu)$ from $\widehat{S}(\nu) = \left|\widehat{f}(\nu)\right|^2$, assuming that some additional information about $f(\nu)$ is available. Now, on the presumption that the complex spectral response function $f(\nu)$ is given by (5.29) and its modulus by (5.33), we can write the error spectrum as

$$E(\nu) = |\beta| \exp[-\mathrm{i}\phi(\nu)], \tag{5.37}$$

where $\phi(\nu)$ is the phase of the error spectrum. Thus the error spectrum has a constant amplitude and only its phase can have a frequency dependence. The error phase $\phi(\nu)$ is now the only quantity in (5.33) that cannot be obtained from the measured modulus $|f(\nu)|$. Consequently, using the MEM in phase retrieval reduces the problem of finding the phase of $f(\nu) = |f(\nu)| \exp[\mathrm{i}\varphi(\nu)]$ to the problem of finding the corresponding error phase.

In order to find a reasonable estimate for $\phi(\nu)$, we must have some additional information about $f(\nu)$. For example, if we know the value of the real (or imaginary) part of $f(\nu)$ or the value of the phase $\varphi(\nu)$ at $L+1$ discrete frequency values ν_l inside the measurement range, we can compute the corresponding error phase values $\phi(\nu_l)$ and estimate the error phase by a polynomial interpolation as

$$\phi(\nu) = B_0 + B_l\nu + ... + B_L\nu^L = \sum_{l=0}^{L} B_l\nu^l, \tag{5.38}$$

where the B_ls satisfy a Vandermonde system [102] as follows:

$$\begin{pmatrix} 1 & \nu_0 & \ldots & \nu_0^L \\ 1 & \nu_1 & \ldots & \nu_1^L \\ \vdots & \vdots & & \vdots \\ 1 & \nu_L & \cdots & \nu_L^L \end{pmatrix} \begin{pmatrix} B_0 \\ B_1 \\ \vdots \\ B_L \end{pmatrix} = \begin{pmatrix} \phi(\nu_0) \\ \phi(\nu_1) \\ \vdots \\ \phi(\nu_L) \end{pmatrix}. \tag{5.39}$$

The idea of using the MEM in obtaining $f(\nu)$ from its modulus is that the error phase is usually a much smoother function than the actual phase and, therefore, that it can be estimated with a low-order polynomial (i.e. the number of required known values of $\phi(\nu_l)$ is low). Clearly, the ideal situation would be that the error phase could be given by a linear estimation i.e. $L \leq 1$ in (5.38). In fact, it is possible to reduce the optimum degree L_{opt} of the polynomial in (5.38) by a simple "squeezing" procedure. Namely, instead of using the measured power spectrum $S(\nu)$ as such in (5.29), we should use a modified spectrum S_{sq} given by

$$S_{\text{sq}}(\nu; K) \equiv \begin{cases} S(\omega_1), & 0 \leq \nu < z_K(\omega_1) \\ S(\omega), & z_K(\omega_1) \leq \nu \leq z_K(\omega_2) \\ S(\omega_2), & z_K(\omega_2) < \nu \leq 1, \end{cases} \tag{5.40}$$

where

$$z_K(\omega) = (2K+1)^{-1}\left(\frac{\omega - \omega_1}{\omega_2 - \omega_1} + K\right) \tag{5.41}$$

and

$$\nu = \frac{z_K(\omega) - z_K(\omega_1)}{z_K(\omega_2) - z_K(\omega_1)}. \tag{5.42}$$

Now, if we choose the squeezing parameter $K = 0$, (5.40) gives the original spectrum as a function of the normalized frequency ν as S_{sq}. However, when $K > 0$, the original is transformed (squeezed) into a narrower range. For example, by setting $K = 2$, the original spectrum $S(\nu)$ is squeezed from the interval $\nu \in [0,1]$ into the interval $\nu \in [0.4, 0.6]$. The rest of the squeezed spectrum is obtained by adding constant wings having the same values as the original spectrum at its end points $\nu = 0$ and $\nu = 1$. Quite commonly, this procedure reduces L_{opt} enough that a linear approximation ($L = 1$) gives a reasonably good estimate of the error phase.

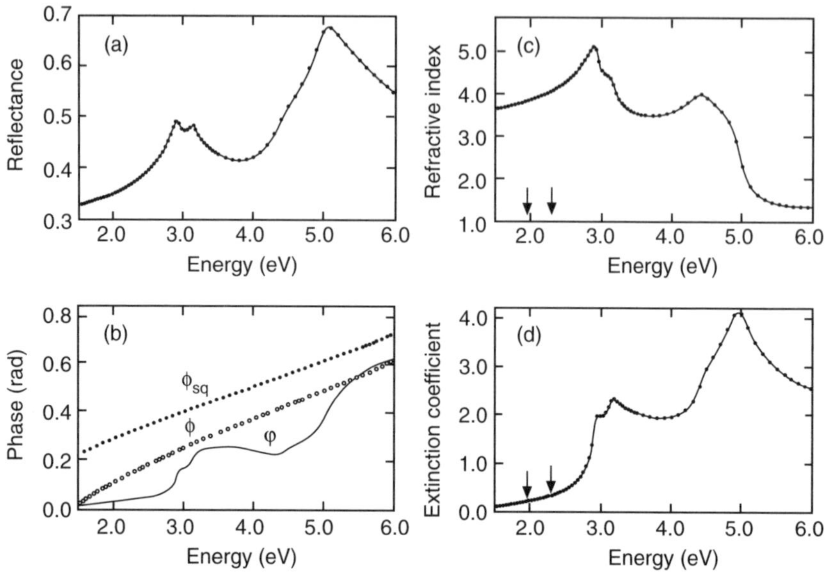

Fig. 5.2. **(a)** Reflectance of GaAs as a function of energy. **(b)** Phase φ and the error phases obtained with (ϕ_{sq}) and without (ϕ) the squeezing procedure. **(c)** Real refractive index, **(d)** extinction coefficient. In (c) and (d), the exact values are shown by *dots* and the computed values by *solid lines*. The *arrows* in (c) and (d) indicate the energies at which the phase values were assumed to be known a priori; e.g. determined by ellipsometry [103]

5.3 Applications in Reflection Spectroscopy

An example of the utility of the maximum entropy model in phase retrieval in reflection spectroscopy is illustrated in Fig. 5.2 [103]. In Fig. 5.2a the reflectance spectrum $R(\omega) = |r(\omega)|^2$ of GaAs is shown within the energy range from 1.5 eV to 6.0 eV. This spectrum (dots) was obtained using complex refractive index data from the literature [104]. The corresponding phase function, $\varphi(\omega)$, and the error phase functions are shown in Fig. 5.2b. Clearly, both of the error phases, obtained with and without the squeezing procedure (as defined in (5.40)–(5.42)), are "slowly varying" functions if compared with $\varphi(\omega)$. Moreover, $\phi_{\mathrm{sq}}(\omega)$ can be well described by a linear approximation. Consequently, the complex refractive index $N(\omega) = n(\omega) + \mathrm{i}k(\omega) = [1 + r(\omega)] / [1 - r(\omega)]$ can be retrieved, if in addition to $R(\omega)$ the value of $r(\omega)$ is known at two frequencies inside the measurement range of $R(\omega)$. In the present example this additional information was assumed to have been obtained by ellipsometric measurements at the wavelengths of red ($\lambda = 633\,\mathrm{nm}$) and green ($\lambda = 543\,\mathrm{nm}$) HeNe lasers. The resultant real (refractive index $n(\omega)$) and imaginary (extinction coefficient $k(\omega)$) parts of $N(\omega)$ are shown

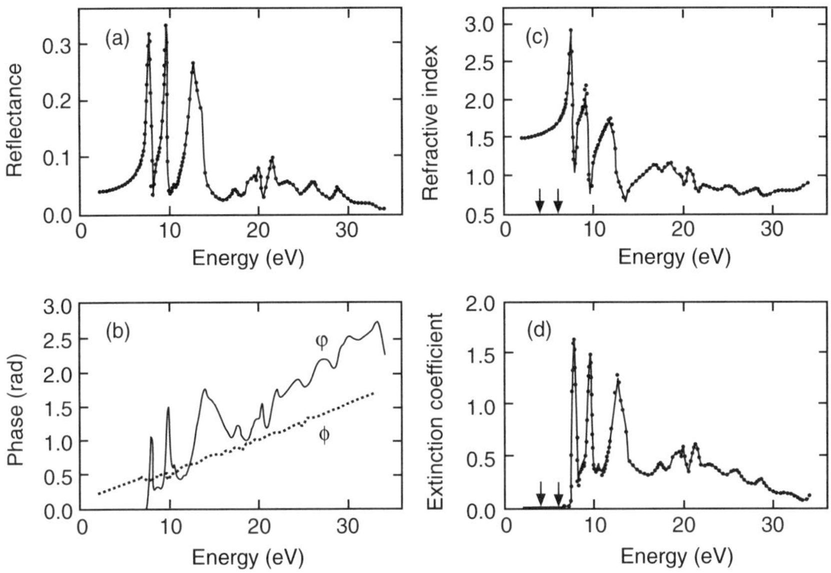

Fig. 5.3. **(a)** Reflectance of a KCl crystal. **(b)** Phase φ and the corresponding error phase ϕ obtained with the squeezing procedure. **(c)** Real refractive index, **(d)** extinction coefficient. In (c) and (d), the exact values are shown by *dots* and the computed values by *solid lines*. The *arrows* in (c) and (d) indicate the energies in the visible range where the extinction coefficient $k \approx 0$; these two points were used to give a linear estimate for the error phase [103]

in Figs. 5.2c,d (solid lines). The exact values (dots) are also shown for comparison.

Another example, concerning the reflectance spectrum of KCl, is shown in Fig. 5.3 [103]. The spectrum shown in Fig. 5.3a extends over energies from the visible (2.0 eV) to the far ultraviolet (34 eV) regions. This spectrum was also obtained using complex refractive index data from the literature [105]. In Fig. 5.3b, the idea of the MEM procedure becomes evident: although the phase $\varphi(\omega)$ has a very complex line shape, the corresponding error phase $\phi_{\mathrm{sq}}(\omega)$ can still be given rather well by a linear approximation. Furthermore, in this case no extra measurement for recovering $\varphi(\omega)$ is needed, because KCl is transparent in the visible range and thus we have the constraint $k(\omega) \approx 0$ when $\hbar\omega \in [2.0\,\mathrm{eV}, 3.1\,\mathrm{eV}]$. Using this constraint, the error phase was estimated and the complex electric-field reflectance was obtained. The resulting real refractive index and extinction coefficient are shown in Figs. 5.3c,d (solid lines). The MEM procedure can also be applied in phase retrieval from magnetoreflectivity data. Such a procedure may have importance in the development of technological devices based on magneto-optics.

5.4 Applications in Nonlinear Optical Spectroscopy

In nonlinear optical spectroscopy the most typical measurement is a power spectrum measurement: the squared modulus $|\chi^{(n)}|^2$ of the nth-order susceptibility $\chi^{(n)} = |\chi^{(n)}| \exp(\mathrm{i}\varphi)$ is measured within some frequency interval. Since both the modulus $|\chi^{(n)}|$ and the phase φ provide crucial information on the multiphoton processes of a medium, the phase retrieval problem arises. It has been shown recently [66] that the phase of the third-order susceptibility $\chi^{(3)}(\omega, \omega, \omega)$ of polydihexylsilane (PDHS) can be computed from its modulus by a Kramers–Kronig integration. In that calculation the extrapolation problem was avoided by a procedure similar to the Roesler method [44]. In the Roesler method, the integrals outside the measurement region were replaced by two experimental phase data points. In Fig. 5.4 these results are shown, together with the corresponding maximum-entropy calculations (the real and imaginary parts of $\chi^{(3)}(\omega, \omega, \omega)$) [106]. The MEM curves were computed by estimating the error phase with first-order ($L = 1$) and third-order ($L = 3$) polynomials. It is observed that these MEM curves are identical, although the curves with $L = 1$ were obtained using only two measured phase values as additional information, whereas all the measured phase values (about 60 values) were used in the cases of the curves with $L = 3$. Moreover, the Kramers–Kronig and the MEM estimates are very similar, except for the fact that the Kramers–Kronig computation could not reproduce the two-photon

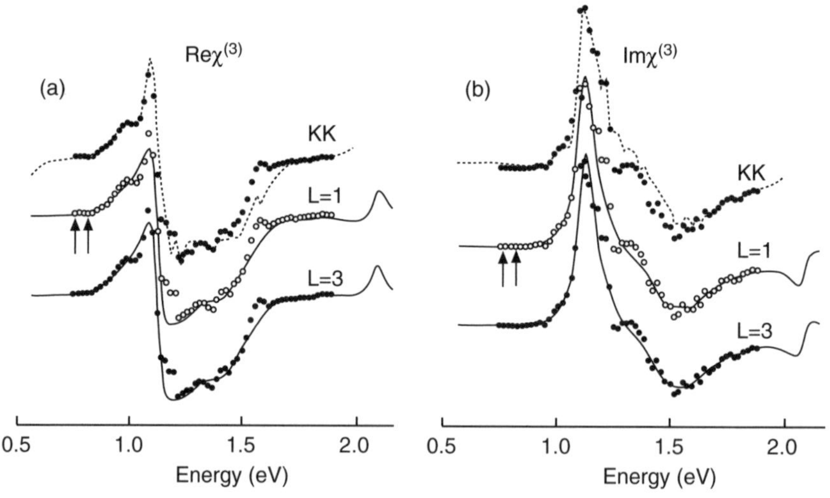

Fig. 5.4. Experimental values (*dots* and *open circles*) of **(a)** real and **(b)** imaginary parts of the susceptibility $\chi^{(3)}(\omega, \omega, \omega)$ of PDHS, and the corresponding curves obtained by a Kramers–Kronig analysis (*dotted lines*) and by the MEM procedure (*solid lines*). The additional information used for the MEM estimates with $L = 1$ was the two phase values indicated by the *arrows* [106]

resonance peak at 2.1 eV. Accordingly, both the Kramers–Kronig and the MEM approaches for phase retrieval required phase data at two frequency points as additional information. However, the significant difference between the Kramers–Kronig and the MEM analyses is the fact that in the Kramers–Kronig calculation the additional information is needed outside, whereas in the MEM it is needed inside the measurement range of $\left|\chi^{(3)}\right|$.

Another third-order susceptibility having much practical interest is the frequency-degenerate susceptibility $\chi^{(3)}(\omega, \omega, -\omega)$, which determines the degenerate four-wave mixing processes. Furthermore, it links together the Kerr coefficient $\gamma(\omega) \propto \operatorname{Re}\chi^{(3)}(\omega, \omega, -\omega)$ and the two-photon absorption coefficient $\beta(\omega) \propto \operatorname{Im}\chi^{(3)}(\omega, \omega, -\omega)$. Unfortunately, the phase of $\chi^{(3)}(\omega, \omega, -\omega)$ cannot be obtained by a Kramers–Kronig integration. This is because the dispersion relation for $\chi^{(3)}(\omega, \omega, -\omega)$, given in (3.24), includes in addition to the ordinary Kramers–Kronig dispersion relation a term due to the contribution of residues. This term can usually be neither computed nor omitted [77].

However, it is possible to use the MEM approach in this case to compute the phase. This has been demonstrated [77] by using a theoretical model for $\chi^{(3)}(\omega, \omega, -\omega)$ (Fig 5.5). Figures 5.5b,c show the real and imaginary parts computed using the amplitude spectrum $\left|\chi^{(3)}(\omega, \omega, -\omega)\right|$ (Fig. 5.5a) with the aid of the MEM procedure and by a Kramers–Kronig integrations (the residue term is omitted). The MEM curves were obtained by assuming that, in addition to the amplitude spectrum, the phase was known at the resonant frequency ω_0. The same assumption was used for Kramers–Kronig integrations apart from the fact that $\left|\chi^{(3)}\right|$ was perfectly "extrapolated" beyond the "measurement" range. The subtractive Kramers–Kronig procedure [37] was used, when use could be made of the fact that the phase was known a priori at a single frequency ω_0. This offers a greater convergence than does the conventional Kramers–Kronig method. Nevertheless, although this procedure forces the phase to have the correct value at ω_0, a good estimate of the dispersion could not be obtained because of the significant contribution of unknown residues. By contrast, the MEM procedure yields a good result.

Before closing, we consider two examples related to meromorphic nonlinear susceptibilities that possess, in addition to the poles, zeros in the upper half of the complex angular frequency plane. One such susceptibility has been studied to some extent already in Sect. 3.5. Indeed the total susceptibility, which includes the information about the linear and nonlinear parts of (3.21), has one zero and one pole in the upper half plane (see (3.22)). In this case the error phase function becomes nonlinear as the ratio I/I_s increases, as shown in Fig. 5.6a. In the case of a small intensity ratio, linear estimation ($L + 1 = 2$) is enough and a priori knowledge of, for example, two values of $\operatorname{Im}\{\chi\}$ is sufficient for good phase retrieval result. In the case of a high intensity ratio, more phase data has to be known. We show in Figs. 5.6b,c the real and imaginary parts obtained by the maximum entropy model in

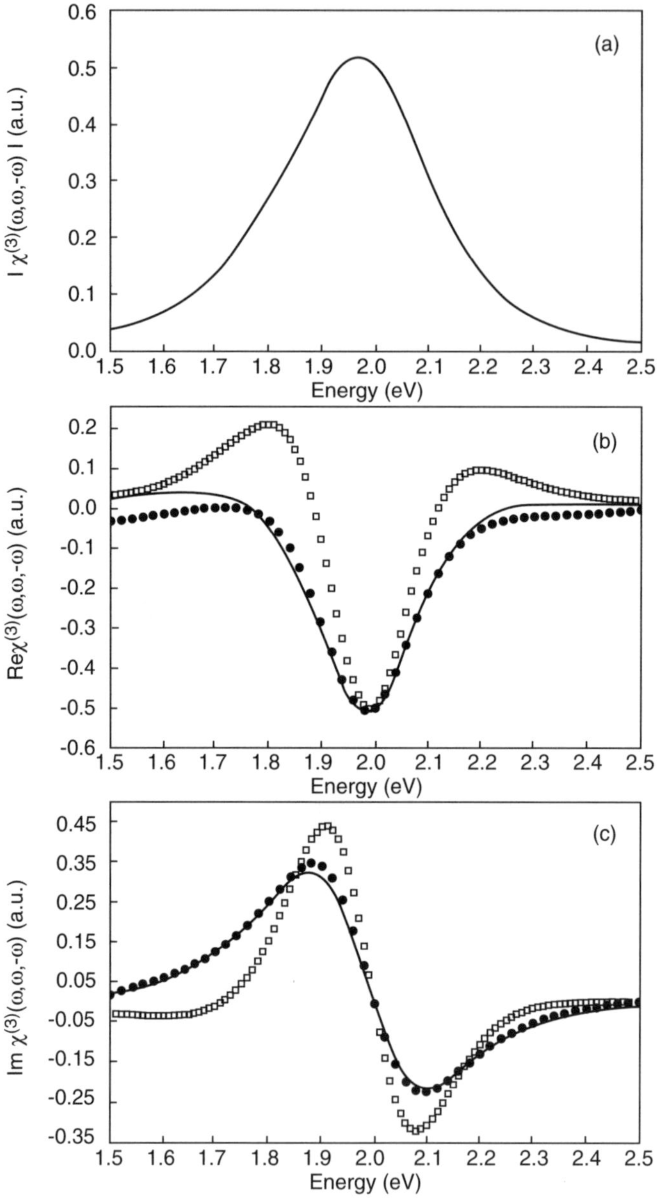

Fig. 5.5. **(a)** Theoretical model for the amplitude $\left|\chi^{(3)}(\omega,\omega,-\omega)\right|$. Kramers–Kronig (*open squares*) and MEM (*filled circles*) estimates of **(b)** $\mathrm{Re}\,\chi^{(3)}$ and **(c)** $\mathrm{Im}\,\chi^{(3)}$ computed from $\left|\chi^{(3)}\right|$. The corresponding *solid lines* show the actual values [77]

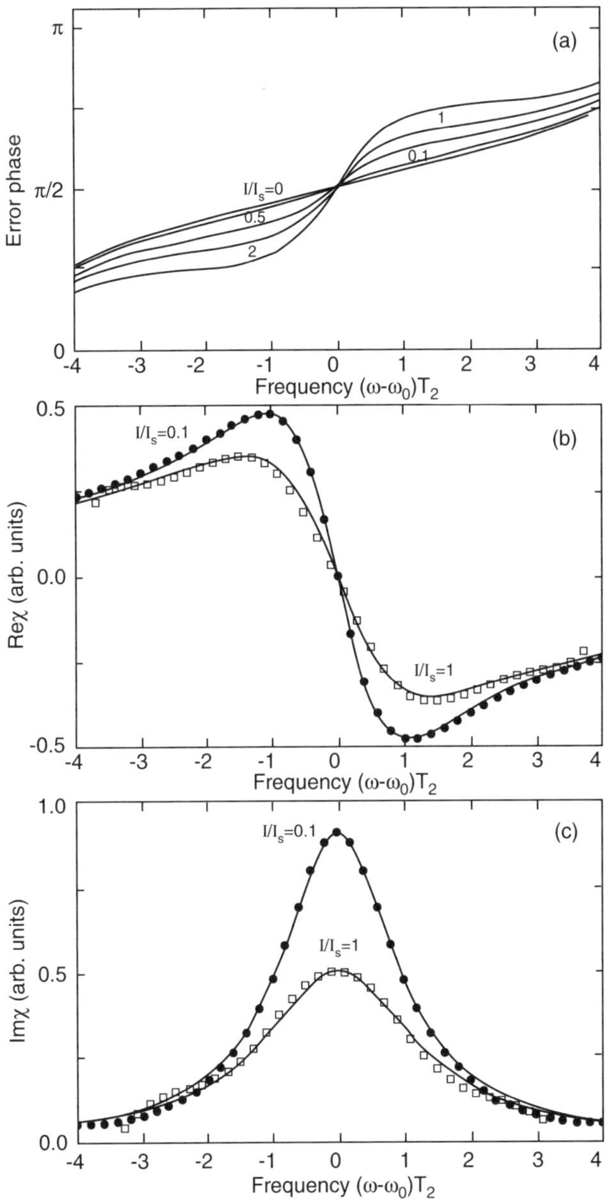

Fig. 5.6. (a) Error phase functions of total susceptibility spectra $|\chi(\omega; I)|^2$ with intensity ratios $I/I_s = 0,\ 0.1,\ 0.5,\ 1.0$. **(b)**, **(c)**, recovered $\mathrm{Re}\,\chi$ and $\mathrm{Im}\,\chi$ curves in cases of low ($I/I_s = 0.1$) and high ($I/I_s = 1.0$) intensity ratios [107]

the cases $I/I_s = 0.1$ ($L = 1$) and $I/I_s = 1.0$ ($L = 5$). Since here the intensity ratio does not affect the actual phase function, the phase retrieval is best performed by measuring a low-intensity spectrum in addition to the high-intensity spectrum, and by computing the phase using the low-intensity data [107].

The other example is related to the nonlinear optical properties of nanocomposites. It is possible to get enhanced nonlinear signals from nanocomposites [45–50]. The physical phenomenon behind the enhancement can be explained in the following manner. In a two-phase nanocomposite, either or both of the host and the inclusions may be nonlinear. With the appropriate choice of the linear permittivities of the two constituents, it is possible to enhance the local electric field in the vicinity of an inclusion. This in turn makes it possible to have enhanced nonlinear optical signals due to the various possible nonlinear processes [108]. As an example, we mention here the nonlinear meromorphic effective susceptibility of Maxwell–Garnett materials, given by [47]

$$\chi_{\text{eff}}^{(3)}(\omega,\omega,-\omega) = f_1 \left| \frac{\varepsilon_{\text{eff}}(\omega) + 2\varepsilon_2(\omega)}{\varepsilon_1(\omega) + 2\varepsilon_2(\omega)} \right|^2 \left[\frac{\varepsilon_{\text{eff}}(\omega) + 2\varepsilon_2(\omega)}{\varepsilon_1(\omega) + 2\varepsilon_2(\omega)} \right]^2 \chi_1^{(3)}(\omega,\omega,-\omega), \quad (5.43)$$

where for simplicity the inclusion particles are taken to consist of a nonlinear material and the host material to be linear. Now the nonlinear susceptibility $\chi_1^{(3)}$ is meromorphic but the squared modulus, involving the permittivities, also introduces meromorphism. In addition, complex zeros appear in the upper half plane. Nevertheless, the maximum entropy model can be applied to resolve the real and imaginary parts of the effective nonlinear susceptibility. This is demonstrated in Fig. 5.7, which was calculated by using a Lorentzian linear permittivity with a single resonance frequency and the meromorphic nonlinear susceptibility of (3.25) for the inclusions.

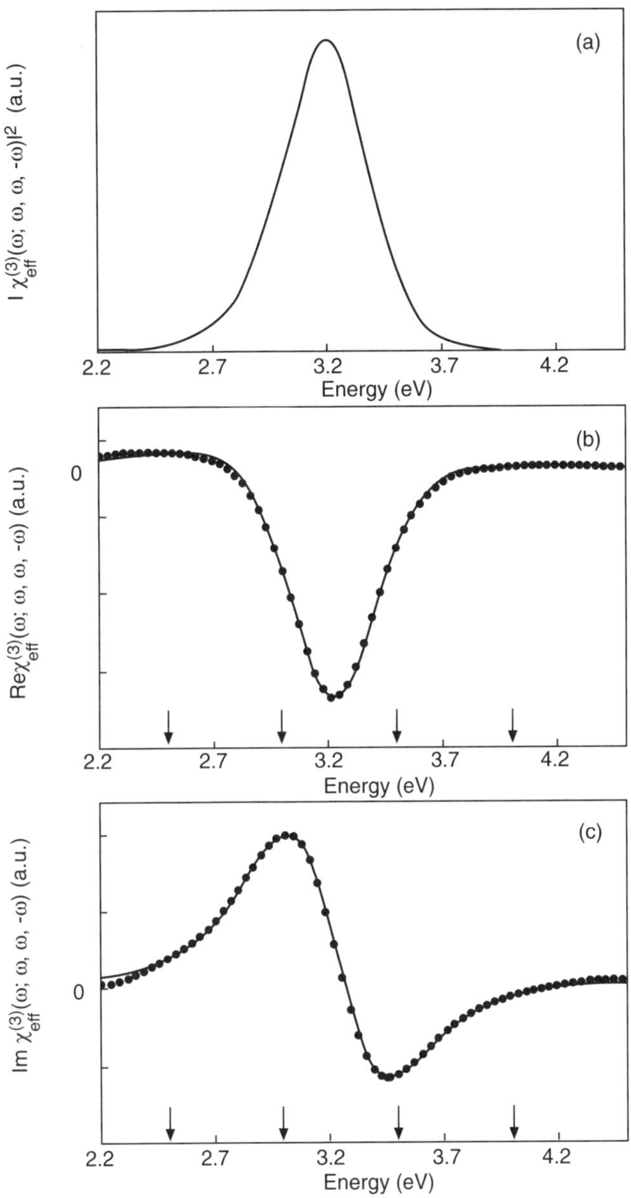

Fig. 5.7. **(a)** Squared modulus of effective nonlinear susceptibility of Maxwell–Garnett material. **(b)**, **(c)** Real and imaginary parts obtained by the MEM procedure, where additional information about the phase is assumed to be known for four equispaced angular frequencies (*arrows*)

6. Sum Rules

Sum rules are important in quantum mechanics, where they yield information about the electronic transitions. To derive these sum rules we usually have to solve the Schrödinger equation to find the wave functions of the states and thereafter calculate, for instance, the dipole matrix elements to obtain the oscillator strengths. After some commutator algebra, we can derive, for example, the Thomas–Reiche–Kuhn sum rule for the oscillator strengths [1]. It is interesting to note that the Kramers–Kronig relations were first derived for the purposes of optics but later adopted in high-energy physics. On the other hand, sum rules have played an important role in high-energy physics for a considerabe time. Superconvergence sum rules for particle physics as described by De Alfaro et al. [109] probably stimulated the study of novel sum rules related to optical constants and nonlinear susceptibilities. In most cases sum rules are used to provide information about the success of data inversion, by either Kramers–Kronig or other methods.

We begin this chapter by dealing with sum rules in linear optical spectroscopy, using complex analysis for their derivations. After that we derive sum rules for holomorphic and meromorphic nonlinear susceptibilities and finally study a mathematical method of finding the numbers of poles and zeros of meromorphic nonlinear susceptibilties.

6.1 *f* Sum Rule and Average-Optical Constant Sum Rules

We start by considering classical sum rules in quantum mechanics and in optical spectroscopy. Actually, quantum mechanics is not at all needed in the derivation of sum rules in optical spectroscopy. Sum rules result from the mathematical properties of the optical constants and nonlinear susceptibilities, which are governed to a great extent by the principle of causality and asymptotic fall-off at high angular frequencies.

In quantum mechanics the oscillator strength is defined by

$$f_{jl} = \frac{2m}{\hbar}\omega_{jl}\left|x_{jl}\right|^2 , \tag{6.1}$$

where x_{jl} is the dipole matrix element between two atomic states j and l. Then the Thomas–Reiche–Kuhn sum rule (a detailed derivation can be found in the book by Wooten [1]) states that

$$\sum_j f_{jl} = Z, \tag{6.2}$$

where Z is the total number of electrons in the atomic system. If the electron density in a medium is $\aleph$ and the oscillator strength density is considered as a function of the angular frequency of the incident photon, then the sum rule (6.2) can be formulated as

$$\int_0^\infty f(\omega)\, \mathrm{d}\omega = \aleph, \tag{6.3}$$

when the system is in its ground state. It is also possible to employ the Kramers–Kronig relation for the linear susceptibility as follows:

$$\mathrm{Re}\left\{\chi^{(1)}(\omega')\right\} = \frac{e^2}{m\varepsilon_0} \mathrm{P} \int_0^\infty \frac{f(\omega)}{\omega^2 - \omega'^2} \mathrm{d}\omega, \tag{6.4}$$

where $f(\omega)\,\mathrm{d}\omega$ is called the number of dispersion electrons in the differential angular frequency range $\mathrm{d}\omega$. Now we find that

$$f(\omega) = \frac{2m\varepsilon_0}{\pi e^2} \omega \,\mathrm{Im}\left\{\chi^{(1)}(\omega)\right\}. \tag{6.5}$$

Using (6.5), we can rewrite (6.3) and find the so-called f sum rule

$$\int_0^\infty \omega \,\mathrm{Im}\left\{\varepsilon(\omega)\right\} \mathrm{d}\omega = \frac{\pi}{2}\frac{\aleph e^2}{m} = \frac{\pi}{2}\omega_\mathrm{p}^2. \tag{6.6}$$

This sum rule can also be deduced from the asymptotic behavior of the linear susceptibility and the Kramers–Kronig relation. To show this, we estimate the complex linear susceptibility at high frequencies in the following manner:

$$\chi^{(1)}(\omega) = -\frac{\omega_\mathrm{p}^2}{\omega^2} + \cdots. \tag{6.7}$$

The appropriate Kramers–Kronig relation for an insulator (metals can be treated in a similar manner but taking account of the DC conductivity) can in turn be estimated as

$$\mathrm{Re}\left\{\chi^{(1)}(\omega')\right\} \to -\frac{2}{\pi}\frac{1}{\omega'^2} \int_0^\infty \omega \,\mathrm{Im}\left\{\chi^{(1)}(\omega)\right\} \mathrm{d}\omega, \;\omega' \to \infty. \tag{6.8}$$

Comparision of (6.7) and (6.8) provides the same information as (6.6). A similar limiting process provides another sum rule, which was discovered by Altarelli et al. [110] using the so-called superconvergence theorem (Appendix

I). The idea of the superconvergence theorem is that in the context of complex contour integration, a strong enough asymptotic fall-off of the optical parameter is guaranteed so that the result of Jordan's lemma is valid. In other words, superconvergence means that the integral along a semicircle vanishes at infinity (the property of superconvergence is estimated with the aid of the Phragmen–Lindelöf theorem of complex functions [10]). Fortunately, the property of superconvergence is present in the best-known cases in linear and nonlinear optical spectroscopy. In the case of the Lorentzian linear susceptibility we can estimate, for high frequencies, that

$$\chi^{(1)}(\widehat{\omega}) \simeq -\frac{\omega_{\mathrm{p}}^2}{\widehat{\omega}^2}\sum_{j=0}^{\infty}\left(\frac{\mathrm{i}\Gamma}{\widehat{\omega}}\right)^j, \tag{6.9}$$

if $|\mathrm{i}\Gamma/\widehat{\omega}| < 1$. We observe then that the imaginary part of the linear susceptibility is proportional to ω^{-3} for real angular frequency. We now proceed and state the result of Altarelli et al. [110]. We have the result that

$$\mathrm{Im}\left\{\chi^{(1)}(\omega')\right\} \to \frac{2}{\pi}\frac{1}{\omega'}\int_0^{\infty}\mathrm{Re}\left\{\chi^{(1)}(\omega)\right\}\mathrm{d}\omega,\ \omega' \to \infty. \tag{6.10}$$

Owing to the asymptotic fall-off of $\mathrm{Im}\left\{\chi^{(1)}(\omega')\right\}$ as in (6.9), the only possibility is that

$$\int_0^{\infty}\mathrm{Re}\left\{\chi^{(1)}(\omega)\right\}\mathrm{d}\omega = 0, \tag{6.11}$$

which states that the average of the real part of the linear susceptibility is zero. Hence $\mathrm{Re}\left\{\chi^{(1)}\right\}$ must have at least one zero on the real axis. Similar arguments to those above and the estimate for the complex refractive index

$$N(\omega) = 1 - \frac{1}{2}\left(\frac{\omega_{\mathrm{p}}^2}{\omega^2}\right) + \cdots \tag{6.12}$$

make it possible to find two sum rules that hold for the extinction coefficient and the real refractive index, namely

$$\int_0^{\infty}\omega k(\omega)\,\mathrm{d}\omega = \frac{\pi}{4}\omega_{\mathrm{p}}^2,$$

$$\int_0^{\infty}[n(\omega) - 1]\,\mathrm{d}\omega = 0. \tag{6.13}$$

Again we can predict that the real refractive index must have at least one zero on the real axis at a positive angular frequency. The numerical approach of using the average sum rule (6.13) in data inversion of experimental data using the Kramers–Kronig dispersion relation is described in the paper of

Altarelli and Smith [111]. In particular, these authors used a verification parameter, related to the average-index sum rule, to show inconsistency in $k(\omega)$ of aluminum. Note that the average-index sum rule (6.13) can also be obtained from (2.42) by changing the variable $\mathrm{i}\omega$ to ω and remembering that $k(\mathrm{i}\omega) \equiv 0$.

We generalize the concept of sum rules to the powers of the complex refractive index and reflectance in the following sections.

6.2 Sum Rules for the Powers of the Complex Refractive Index

We have already shown that the complex refractive index is a holomorphic function in the upper half plane. Powers of holomorphic functions are also holomorphic functions, and in addition, products of powers of the complex variable and powers of holomorphic functions are holomorphic functions. This observation stimulated Altarelli and Smith [111] to derive a set of new sum rules with regard to the complex refractive index. Nevertheless, their derivation was based to a great extent on the use of Kramers–Kronig-type relations and, furthermore, on the appropriate asymptotic fall-off of functions at low and high angular frequencies. Actually, an appropriate choice of the powers can enhance the convergence of the intergrals, and the problem of data extrapolation, which is always present in connection with sum rules, becomes less significant.

Let us study the functions $[N(\hat{\omega})-1]^m$ and $\hat{\omega}^m [N(\hat{\omega})-1]^m$, where m is a positive integer, which are holomorphic in the upper half plane. Then we find, using the technique of complex contour intergration, familiar in the context of the derivation of the Kramers–Kronig relations, relations such as

$$\mathrm{Re}\left\{[N(\omega')-1]^m\right\} = \frac{2}{\pi}\mathrm{P}\int_0^\infty \frac{\omega\,\mathrm{Im}\left\{[N(\omega)-1]^m\right\}}{\omega^2-\omega'^2}\mathrm{d}\omega\,,$$

$$\mathrm{Im}\left\{[N(\omega')-1]^m\right\} = -\frac{2\omega'}{\pi}\mathrm{P}\int_0^\infty \frac{\mathrm{Re}\left\{[N(\omega)-1]^m\right\}}{\omega^2-\omega'^2}\mathrm{d}\omega\,, \tag{6.14}$$

$$\omega'^m\,\mathrm{Re}\left\{[N(\omega')-1]^m\right\} = \begin{cases} \dfrac{2\omega'}{\pi}\mathrm{P}\displaystyle\int_0^\infty \frac{\omega^m\,\mathrm{Im}\left\{[N(\omega)-1]^m\right\}}{\omega^2-\omega'^2}\mathrm{d}\omega, & m \text{ odd} \\[2ex] \dfrac{2}{\pi}\mathrm{P}\displaystyle\int_0^\infty \frac{\omega^{m+1}\,\mathrm{Re}\left\{[N(\omega)-1]^m\right\}}{\omega^2-\omega'^2}\mathrm{d}\omega, & m \text{ even} \end{cases} \tag{6.15}$$

and

$$\omega'^m \operatorname{Im}\left\{\left[N\left(\omega'\right)-1\right]^m\right\} = \begin{cases} -\dfrac{2}{\pi}\mathrm{P}\displaystyle\int_0^\infty \dfrac{\omega^{m+1}\operatorname{Re}\left\{\left[N\left(\omega\right)-1\right]^m\right\}}{\omega^2-\omega'^2}\mathrm{d}\omega, & m \text{ odd} \\ -\dfrac{2\omega'}{\pi}\mathrm{P}\displaystyle\int_0^\infty \dfrac{\omega^{m}\operatorname{Re}\left\{\left[N\left(\omega\right)-1\right]^m\right\}}{\omega^2-\omega'^2}\mathrm{d}\omega, & m \text{ even.} \end{cases} \tag{6.16}$$

The first relation in (6.14) provides us with information about the powers of the static refractive index in the following way:

$$\left[n\left(0\right)-1\right]^m = \frac{2}{\pi}\mathrm{P}\int_0^\infty \frac{\operatorname{Im}\left\{\left[N\left(\omega\right)-1\right]^m\right\}}{\omega}\mathrm{d}\omega. \tag{6.17}$$

Generally speaking, sum rules that involve only either the real refractive index or the extinction coefficient are more useful. The reason is that usually we can measure only either the imaginary or the real part of the complex refractive index. Therefore we wish to use a sum rule that is directly related to the measured quantity. If data inversion is performed by the Kramers–Kronig relations, the success of the data inversion can be estimated by an appropriate sum rule. Such a test is possible also with mixed sum rules. By mixed sum rules we mean a sum rule where the real and imaginary parts of the optical constants appear simultaneously.

If we choose $m = 1$ in (6.16) and let $\omega' \to 0$, we get the second equation of (6.13), which involves solely the change of the real refractive index and expresses the restatement of the principle of causality. In the general case for odd m in (6.16), and after the limiting process $\omega' \to 0$, we obtain a (mixed) sum rule such as

$$\int_0^\infty \omega^{m-1}\operatorname{Re}\left\{\left[N\left(\omega\right)-1\right]^m\right\}\mathrm{d}\omega = 0,\ m = 1,\ 3,\cdots, \tag{6.18}$$

which holds for both insulators and metals.

As we have already noticed, the average sum rule for the real refractive index change means that the function $n\left(\omega\right)-1$ must have at least one zero, which we denote by ω_{z}. Then, for the case of $m = 2$, it follows from (6.14) and (6.16) that

$$-\frac{2\omega_{\mathrm{z}}}{\pi}\mathrm{P}\int_0^\infty \frac{\operatorname{Re}\left\{\left[N\left(\omega\right)-1\right]^2\right\}}{\omega^2-\omega_{\mathrm{z}}^2}\mathrm{d}\omega = 0\,,$$

$$-\frac{2\omega_{\mathrm{z}}}{\pi}\mathrm{P}\int_0^\infty \frac{\omega^2\operatorname{Re}\left\{\left[N\left(\omega\right)-1\right]^2\right\}}{\omega^2-\omega_{\mathrm{z}}^2}\mathrm{d}\omega = 0\,. \tag{6.19}$$

Multiplication of the latter equation by ω_z^2 and subtraction yields [112]

$$\frac{2\omega_z}{\pi} \mathrm{P} \int_0^\infty \frac{\omega^2 \operatorname{Re}\left\{[N(\omega)-1]^2\right\}}{\omega^2-\omega_z^2} \mathrm{d}\omega - \frac{2\omega_z^3}{\pi} \mathrm{P} \int_0^\infty \frac{\operatorname{Re}\left\{[N(\omega)-1]^2\right\}}{\omega^2-\omega_z^2} \mathrm{d}\omega = 0, \tag{6.20}$$

which reduces to a sum rule given by Altarelli et al. [110],

$$\int_0^\infty [n(\omega)-1]^2 \mathrm{d}\omega = \int_0^\infty k^2(\omega) \mathrm{d}\omega. \tag{6.21}$$

Sum rules for the complex refractive index can be derived using different weighting functions [113], Dirac's delta function [40], complex contour integration [114, 115] and a subtractive technique using the zeros as above [40, 112].

Before closing, we can readily find from the Kramers–Kronig dispersion relation that

$$\mathrm{P} \int_0^\infty \frac{\omega k(\omega)}{\omega^2-\omega_z^2} \mathrm{d}\omega = 0\,, \tag{6.22}$$

which provides a means to find zero/zeros of the refractive index function. In addition, the concept of f sum rules can be generalized, as realized by Villani and Zimerman [113]. Indeed, using complex contour integrations, superconvergence theorem and a function defined by

$$f_m(\omega) = \frac{\left\{2\omega^2 [N(\omega)-1]\right\}^m - (-1)^m \omega_\mathrm{p}^{2m}}{\omega}\,, \tag{6.23}$$

Villani and Zimerman [113] found the following generalized f sum rule for the optical constants of nonmagnetic materials:

$$\int_{-\infty}^\infty \omega^{2m-1} [N(\omega)-1]^m \mathrm{d}\omega = \mathrm{i}(-1)^{m+1} 2^{-m} \pi \omega_\mathrm{p}^{2m}, \tag{6.24}$$

where m is a positive integer.

6.3 Sum Rules for the Powers of the Complex Reflectivity

It is not difficult to imagine that sum rules can also be found for the complex reflectivity. Indeed, using either superconvergence [116] or the Dirac's delta function technique [117], we can find various sum rules for the complex reflectance. Unfortunately, the sum rules for the complex reflectivity are not as practical as for the refractive index since they involve mixed sum rules.

Only for an imaginary angular frequency it is possible to find a sum rule that gives information about the amplitude reflectivity [117]. In other cases the amplitude and the phase angle are always coupled together when we write down sum rules. The main problem with the complex reflectivity is that when we take the natural logarithm in order to resolve the modulus and the phase, superconvergence is no longer valid. However, if we consider the powers of the complex reflectivity, superconvergence exists and a set of sum rules can be constructed. This is accomplished by inspection of holomorphic functions such as $r^m(\widehat{\omega})$, $\widehat{\omega}^m r^m(\widehat{\omega})$ and $\widehat{\omega}^{2(m-1)} r^m(\widehat{\omega})$. Then we can write, just as in the case of the complex refractive index, a set of dispersion relations, which are [116]

$$
\begin{aligned}
|r(\omega')|^m \cos[m\varphi(\omega')] &= \frac{2}{\pi}\mathrm{P}\int_0^\infty \frac{\omega\,|r(\omega)|^m \sin[m\varphi(\omega)]}{\omega^2-\omega'^2}\,\mathrm{d}\omega,\\
|r(\omega')|^m \sin[m\varphi(\omega')] &= -\frac{2\omega'}{\pi}\mathrm{P}\int_0^\infty \frac{|r(\omega)|^m \cos[m\varphi(\omega)]}{\omega^2-\omega'^2}\,\mathrm{d}\omega,
\end{aligned}
\tag{6.25}
$$

$$
\omega'^m |r(\omega')|^m \cos[m\varphi(\omega')] = \begin{cases} \dfrac{2\omega'}{\pi}\mathrm{P}\displaystyle\int_0^\infty \frac{\omega^m |r(\omega)|^m \sin[m\varphi(\omega)]}{\omega^2-\omega'^2}\,\mathrm{d}\omega, & m \text{ odd}\\[2ex] \dfrac{2}{\pi}\mathrm{P}\displaystyle\int_0^\infty \frac{\omega^{m+1} |r(\omega)|^m \sin[m\varphi(\omega)]}{\omega^2-\omega'^2}\,\mathrm{d}\omega, & m \text{ even,} \end{cases}
\tag{6.26}
$$

$$
\omega'^m |r(\omega')|^m \sin[m\varphi(\omega')] = \begin{cases} -\dfrac{2}{\pi}\mathrm{P}\displaystyle\int_0^\infty \frac{\omega^{m+1} |r(\omega)|^m \cos[m\varphi(\omega)]}{\omega^2-\omega'^2}\,\mathrm{d}\omega, & m \text{ odd}\\[2ex] -\dfrac{2\omega'}{\pi}\mathrm{P}\displaystyle\int_0^\infty \frac{\omega^{m} |r(\omega)|^m \cos[m\varphi(\omega)]}{\omega^2-\omega'^2}\,\mathrm{d}\omega, & m \text{ even,} \end{cases}
\tag{6.27}
$$

and

$$
\begin{aligned}
\omega'^{2(m-1)} |r(\omega')|^m \cos[m\varphi(\omega')] &= \frac{2}{\pi}\mathrm{P}\int_0^\infty \frac{\omega^{2m-1} |r(\omega)|^m \sin[m\varphi(\omega)]}{\omega^2-\omega'^2}\,\mathrm{d}\omega\,,\\
\omega'^{2(m-1)} |r(\omega')|^m \sin[m\varphi(\omega')] &= -\frac{2\omega'}{\pi}\mathrm{P}\int_0^\infty \frac{\omega^{2(m-1)} |r(\omega)|^m \cos[m\varphi(\omega)]}{\omega^2-\omega'^2}\,\mathrm{d}\omega,
\end{aligned}
\tag{6.28}
$$

where the symmetry properties of (2.44) have been employed. From (6.25) we observe that the static value of the electric-field amplitude reflectivity obeys the sum rule

$$|r(0)|^m = \frac{2}{\pi} \mathrm{P} \int_0^\infty \frac{|r(\omega)|^m \sin[m\varphi(\omega)]}{\omega} \mathrm{d}\omega. \tag{6.29}$$

In the case of insulators the value of the integral in (6.29) is less than unity, whereas for metals, irrespective of the value of the integer m, it will always give exactly unity. Sum rules are obtained by inspection of the low- and high-frequency limits of the complex reflectivity as well as the dispersion relations above. With the aid of (6.26) we find for the low-frequency limit the following sum rules:

$$\int_0^\infty \omega^{m-2} |r(\omega)|^m \sin[m\varphi(\omega)] \mathrm{d}\omega, \; m = 3,\; 5,\; 7, \cdots, \tag{6.30}$$

and

$$\int_0^\infty \omega^{m-1} |r(\omega)|^m \sin[m\varphi(\omega)] \mathrm{d}\omega = 0, \; m = 2,\; 4,\; 6, \cdots. \tag{6.31}$$

In a similar manner, by using the low-frequency limit and (6.27), we can write also the sum rules

$$\int_0^\infty \omega^{m-1} |r(\omega)|^m \cos[m\varphi(\omega)] \mathrm{d}\omega = 0, \; m \text{ odd} \tag{6.32}$$

and

$$\int_0^\infty \omega^{m-2} |r(\omega)|^m \cos[m\varphi(\omega)] \mathrm{d}\omega = 0, \; m \text{ even}. \tag{6.33}$$

The derivation of sum rules in the high-frequency limit is based on the superconvergence of the function in question and therefore on the asymptotic fall-off of the complex reflectivity. The high-frequency limit of the complex reflectivity tends to

$$r(\omega) \to -\frac{1}{4} \frac{\omega_\mathrm{p}^2}{\omega^2}, \text{ as } \omega \to \infty. \tag{6.34}$$

Now, using the information of (6.26), (6.27) and (6.34), we get the following sum rules:

$$\int_0^\infty \omega^m |r(\omega)|^m \sin[m\varphi(\omega)] \mathrm{d}\omega = \begin{cases} \dfrac{\pi\omega_\mathrm{p}^2}{8}, & m = 1 \\ 0, & m = 3,\; 5,\; 7, \cdots, \end{cases} \tag{6.35}$$

$$\int_0^\infty \omega^{m+1} |r(\omega)|^m \cos[m\varphi(\omega)]\, \mathrm{d}\omega = \begin{cases} -\dfrac{\pi\omega_\mathrm{p}^4}{32}, & m = 2 \\ 0, & m = 4,\ 6,\ 8, \cdots, \end{cases} \tag{6.36}$$

$$\int_0^\infty \omega^{m+1} |r(\omega)|^m \cos[m\varphi(\omega)]\, \mathrm{d}\omega = 0,\ m = 3,\ 5,\ 7, \cdots \tag{6.37}$$

and

$$\int_0^\infty \omega^{m} |r(\omega)|^m \cos[m\varphi(\omega)]\, \mathrm{d}\omega = 0,\ m = 2,\ 4,\ 6, \cdots . \tag{6.38}$$

A generalized f sum rule and an inertial sum rule are obtained, with the aid of the high-angular-frequency limit

$$\omega^{2(m-1)} r^m(\omega) \to \left(-\frac{1}{4}\right)^m \frac{\omega_\mathrm{p}^{2m}}{\omega^2},\ \text{as}\ \omega \to \infty , \tag{6.39}$$

and these are given by [116]

$$\int_0^\infty \omega^{2m-1} |r(\omega)|^m \sin[m\varphi(\omega)]\, \mathrm{d}\omega = \frac{(-1)^m \pi\omega_\mathrm{p}^{2m}}{2^{2m+1}} , \tag{6.40}$$

and

$$\int_0^\infty \omega^{2(m-1)} |r(\omega)|^m \cos[m\varphi(\omega)]\, \mathrm{d}\omega = 0. \tag{6.41}$$

Note that the derivative of the complex reflectivity, such as in (4.17), provides also a means to find sum rules. Indeed, the function $(\mathrm{d}r/\mathrm{d}\omega)/r$ is holomorphic in the upper half plane and, therefore, possesses the asymptotic fall-off needed in Jordan's lemma. After applying the symmetry properties we can write the following dispersion relations for insulators:

$$\frac{\mathrm{d}r(\omega')/\mathrm{d}\omega'}{r(\omega')} = \frac{2\omega'}{\pi} \mathrm{P} \int_0^\infty \frac{\mathrm{d}\varphi(\omega)/\mathrm{d}\omega}{\omega^2 - \omega'^2} \mathrm{d}\omega ,$$

$$\frac{\mathrm{d}\varphi(\omega')}{\mathrm{d}\omega'} = -\frac{2}{\pi} \mathrm{P} \int_0^\infty \frac{\omega\{[\mathrm{d}r(\omega)/\mathrm{d}\omega]/r(\omega)\}}{\omega^2 - \omega'^2} \mathrm{d}\omega. \tag{6.42}$$

We immediately find, in the low-angular-frequency limit, that

$$\left.\frac{\mathrm{d}\varphi}{\mathrm{d}\omega'}\right|_{\omega'=0} = -\frac{2}{\pi} \mathrm{P} \int_0^\infty \frac{\mathrm{d}r(\omega)/\mathrm{d}\omega}{\omega r(\omega)} \mathrm{d}\omega. \tag{6.43}$$

Finally we remark that since there is an indeterminacy such that the same phase is associated with electric-field reflectivities $|r|$ and $C|r|$, where C is a constant and $C|r| < 1$, we cannot derive conventional f sum rules separately for the phase and the modulus. King [117] considered imaginary angular frequencies and could derive a sum rule related to $R = rr^*$. It can be written in the following form:

$$\int_0^\infty \frac{\ln R(\omega)}{\omega^2 + \omega'^2} \mathrm{d}\omega = \frac{\pi}{2\omega'} \ln R(\mathrm{i}\omega'), \ \omega' \neq 0. \tag{6.44}$$

The main benefit of this sum rule is that it can help us in checking the consistency of theoretical models for the reflectivity of a medium.

6.4 Sum Rules in Magneto-Optics

Sum rules for the complex refractive index and reflectivity in magneto-optics follow in a similar manner described above. That is to say, the complex contour integration of holomorphic functions is applied. Just as in the previous sections, low- and high-angular frequency limits together with superconvergence provide us with some typical sum rules. Here we present some of them for the complex refractive index and reflectance in the case of the Faraday configuration [53, 54, 118].

In the case of insulators, for zero angular frequency,

$$\begin{aligned} n_+(0) &= n_-(0) = n(0) \ , \\ k_+(0) &= k_-(0) = (0) \ . \end{aligned} \tag{6.45}$$

Then we obtain directly from (2.70) a sum rule at the static limit:

$$n(0) - 1 = \frac{1}{\pi} \mathrm{P} \int_0^\infty \frac{k_+(\omega) + k_-(\omega)}{\omega} \mathrm{d}\omega. \tag{6.46}$$

Another sum rule, obtained by inspection of (2.71), is

$$\int_0^\infty \frac{n_+(\omega) - n_-(\omega)}{\omega} \mathrm{d}\omega = 0. \tag{6.47}$$

Two other sum rules for the low-frequency limit can be obtained by inspection of $\widehat{\omega}[N_\pm(\widehat{\omega}) - 1]$, which is a holomorphic function in the upper half plane. Indeed, we can write dispersion relations such as

$$\frac{\omega'[k_+(\omega') + k_-(\omega')]}{2} = -\frac{2}{\pi} \mathrm{P} \int_0^\infty \frac{\omega^2}{\omega^2 - \omega'^2} \left[\frac{n_+(\omega) + n_-(\omega)}{2} - 1\right] \mathrm{d}\omega \ ,$$

$$\omega'[n_+(\omega') - n_-(\omega')] = \frac{2}{\pi} \mathrm{P} \int_0^\infty \frac{\omega^2}{\omega^2 - \omega'^2} [k_+(\omega) - k_-(\omega)] \mathrm{d}\omega \ , \tag{6.48}$$

and then let $\omega' \to 0$, which implies the sum rules

$$\int_0^\infty \left[\frac{n_+(\omega) + n_-(\omega)}{2} - 1\right] \mathrm{d}\omega = 0 \tag{6.49}$$

and

$$\int_0^\infty [k_+(\omega) - k_-(\omega)]\, \mathrm{d}\omega = 0, \tag{6.50}$$

which hold for both insulators and metals.

At the high-angular-frequency limit and exploiting the superconvergence property, we find with the aid of the first formula in (2.70) that

$$\int_0^\infty \omega \frac{k_+(\omega) + k_-(\omega)}{2} \mathrm{d}\omega = \frac{\pi}{4}\omega_\mathrm{p}^2. \tag{6.51}$$

In a similar manner, by inspection of the second formula in (2.71), it follows that

$$\int_0^\infty \omega\, [n_+(\omega) - n_-(\omega)]\, \mathrm{d}\omega = 0. \tag{6.52}$$

This, in turn, along with the definition of (2.72), means that the Faraday rotation angle obeys the sum rule

$$\int_0^\infty \Phi_\mathrm{F}(\omega)\, \mathrm{d}\omega = 0. \tag{6.53}$$

In the case of magnetoreflectivity, we can find sum rules as well [54]. We state here only that since $\varphi_+(0) - \varphi_-(0) = 0$, a sum rule is obtained from (2.75) in the low-frequency limit, for instance, and is given by

$$\int_0^\infty \frac{1}{\omega} \ln \frac{|r_+(\omega)|}{|r_-(\omega)|} \mathrm{d}\omega = 0. \tag{6.54}$$

Other sum rules for magnetoreflectivities are presented in [53, 119] for both the Faraday and the Voigt configurations.

6.5 Sum Rules in Nonlinear Optics

Sum rules for nonlinear susceptibilities can be derived with the aid of complex analysis. If we exploit the theory of several complex variables for holomorphic nonlinear susceptibilities we find that the powers of nonlinear susceptibilities can be holomorphic functions. Also, functions that are obtained by multiplying powers of holomorphic nonlinear susceptibilities by powers of the complex angular frequency constitute holomorphic functions. One example of such a function is $\widehat{\omega}_j^{s-1}\left[\chi^{(n)}\left(\widehat{\omega}_1,...,\widehat{\omega}_j,...,\widehat{\omega}_n\right)\right]^m$, $s=1,2,...,\ m=1,2...,\ s\le m$. We can write dispersion relations for this function and make use of the low- and high-angular-frequency limits as we did in linear optical spectroscopy. However, it is also possible to treat the holomorphic function more directly by complex contour integration:

$$
\begin{aligned}
&\int_{-\infty}^{\infty} \omega_j^{s-1}\left[\chi^{(n)}\left(\omega_1,...,\omega_j,...,\omega_n\right)\right]^m \mathrm{d}\omega_j \\
&= \lim_{B\to\infty}\left(\oint_C \widehat{\omega}_j^{s-1}\left[\chi^{(n)}\left(\widehat{\omega}_1,...,\widehat{\omega}_j,...,\widehat{\omega}_n\right)\right]^m \mathrm{d}\widehat{\omega}_j\right. \\
&\qquad \left.-\int_A \widehat{\omega}_j^{s-1}\left[\chi^{(n)}\left(\widehat{\omega}_1,...,\widehat{\omega}_j,...,\widehat{\omega}_n\right)\right]^m \mathrm{d}\widehat{\omega}_j\right),
\end{aligned} \tag{6.55}
$$

where the integrals on the right-hand side are equal to zero according to Cauchy's integral theorem and Jordan's lemma. Then we can find, for instance [115],

$$
\int_{-\infty}^{\infty} \omega_j^{s-1}\left[\chi^{(n)}\left(\omega_1,...,\omega_j,...,\omega_n\right)\right]^m \mathrm{d}\omega_j = 0, \tag{6.56}
$$

provided that the nonlinear susceptibility is not meromorphic. In the case of third-harmonic wave generation we can deduce from (6.56) that

$$
\int_0^{\infty} \mathrm{Re}\left\{\chi^{(3)}\left(\omega,\omega,\omega\right)\right\} \mathrm{d}\omega = 0, \tag{6.57}
$$

where the symmetry property $\chi^{(3)}\left(-\omega,-\omega,-\omega\right)=\left[\chi^{(3)}\left(\omega,\omega,\omega\right)\right]^*$ has been employed. Furthermore, sum rules for meromorphic nonlinear susceptibilities can be obtained if the complex integration is not performed using the angular frequency variable that produces poles in both half planes simultaneously. Then we can write, for instance [120],

$$
\int_{-\infty}^{\infty} \omega_j^{s-1}\left[\chi^{(n)}\left(\omega_1,...,\omega_j,\omega_k,-\omega_k,...,\omega_n\right)\right]^m \mathrm{d}\omega_j = 0. \tag{6.58}
$$

If we scan the angular frequency that produces the meromorphism and keep the other frequencies constant, a sum rule can be written for the general case [77] with the aid of the theorem of residues

$$\int_{-\infty}^{\infty} \chi^{(n)}(\omega_1, ..., \omega_k, -\omega_k, ..., \omega_n)\, d\omega_k$$
$$= 2\pi i \sum_{\mathrm{Im}\,\widehat{\omega}_k > 0} \mathrm{Res}\left[\chi^{(n)}(\omega_1, ..., \widehat{\omega}_k, -\widehat{\omega}_k, ..., \omega_n)\right] . \tag{6.59}$$

It is obvious that sum rules for nonlinear susceptibilities which are analogous to those for the linear optical constants can be constructed.

Sum rules that hold for the nonlinear part of the complex refractive index can also be formulated [68]:

$$\int_0^{\infty} \omega_1 k_{\mathrm{NL}}(\omega_1, \omega_2)\, d\omega_1 = 0 ,$$
$$\int_0^{\infty} n_{\mathrm{NL}}(\omega_1, \omega_2)\, d\omega_1 = 0 . \tag{6.60}$$

Sum rules of the form (6.60) can also be written for the nonlinear parts of the total permittivity. Using an anharmonic Lorentz oscillator model, Scandolo and Bassani [121] calculated the functions $\omega_1 \mathrm{Im}\{\varepsilon_{\mathrm{total}}(\omega_1, \omega_2)\}$ and $\omega_1 \mathrm{Im}\{\varepsilon^{(1)}(\omega_1, \omega_2)\}$, which are illustrated in Fig. 6.1 for the case of a probe and pump field system. The pump field has an effect on the shape of $\omega_1 \mathrm{Im}\{\varepsilon_{\mathrm{total}}(\omega_1, \omega_2)\}$. In Fig. 6.1 the nonlinear part of the function $\omega_1 \mathrm{Im}\{\varepsilon_{\mathrm{NL}}(\omega_1, \omega_2)\} = \omega_1 \mathrm{Im}\left\{\chi^{(3)}(\omega_1, \omega_2)\right\} = \omega_1 \left[\mathrm{Im}\{\varepsilon_{\mathrm{total}}(\omega_1, \omega_2)\} - \mathrm{Im}\left\{\varepsilon^{(1)}(\omega_1, \omega_2)\right\}\right]$ is also shown. This function obeys an average sum rule similar to the first equation in (6.60).

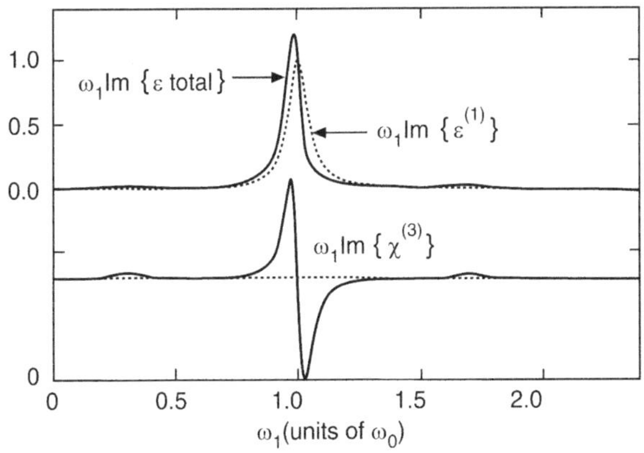

Fig. 6.1. Functions $\omega_1 \mathrm{Im}\{\varepsilon_{\mathrm{total}}(\omega_1, \omega_2)\}$, $\omega_1 \mathrm{Im}\left\{\varepsilon^{(1)}(\omega_1, \omega_2)\right\}$ and $\omega_1 \mathrm{Im}\{\varepsilon_{\mathrm{NL}}(\omega_1, \omega_2)\} = \omega_1 \mathrm{Im}\{\chi^{(3)}(\omega_1, \omega_2)\}$ for an anharmonic Lorentz oscillator [121]

Next we describe in more detail how sum rules can provide a method to estimate the background susceptibility in CARS when the data analysis is based on the maximum entropy model. By measurement, we have information about $\left|\chi^{(3)}(\Delta = \omega_1 - \omega_2)\right|^2 = \left|\chi^{(3)}_{\mathrm{NR}} + \chi^{(3)}_{\mathrm{R}}(\Delta)\right|^2$. Also, we know that the sum rules

$$\int_0^\infty \left[\mathrm{Re}\left\{\chi^{(3)}(\Delta)\right\} - \chi^{(3)}_{\mathrm{NR}}\right] \mathrm{d}\Delta = 0\,,$$

$$\int_0^\infty \left[\mathrm{Re}\left\{\chi^{(3)}(\Delta)\right\} - \chi^{(3)}_{\mathrm{NR}}\right]^2 \mathrm{d}\Delta = \int_0^\infty \left[\mathrm{Im}\left\{\chi^{(3)}(\Delta)\right\}\right]^2 \mathrm{d}\Delta \tag{6.61}$$

are valid for the Raman susceptibility. The sum rules are used in practise in a finite measurement range $\Delta_1 \leq \Delta \leq \Delta_2$. According to the first equation in (6.61) we can write

$$I_1(\Delta_{21}) = \frac{1}{\Delta_{21}} \int_{\Delta_1}^{\Delta_2} \mathrm{Re}\left\{\chi^{(3)}(\Delta)\right\} \mathrm{d}\Delta = \chi^{(3)}_{\mathrm{NR}} \pm |\delta_{21}|\,, \tag{6.62}$$

where $\Delta_{21} = \Delta_2 - \Delta_1$ and $\pm|\delta_{21}|$ is a deviation parameter whose value depends on Δ_{21}; consequently $|\delta_{21}| \to 0$ when $\Delta_{21} \to \infty$. Furthermore, we can assume that when Δ is far away from resonances, $\mathrm{Im}\left\{\chi^{(3)}(\Delta)\right\}$ falls off faster than $\mathrm{Re}\left\{\chi^{(3)}(\Delta)\right\} - \chi^{(3)}_{\mathrm{NR}}$. Thus, it follows from the second equation in (6.61) that

$$\begin{aligned} I_2(\Delta_{21}) &= \int_{\Delta_1}^{\Delta_2} \left[\mathrm{Re}\left\{\chi^{(3)}(\Delta)\right\} - \chi^{(3)}_{\mathrm{NR}}\right] \mathrm{d}\Delta \\ &\leq \int_{\Delta_1}^{\Delta_2} \left[\mathrm{Im}\left\{\chi^{(3)}(\Delta)\right\}^2\right] \mathrm{d}\Delta = I_3(\Delta_{21})\,. \end{aligned} \tag{6.63}$$

Now, we can directly compute the above integrals $I_2(\Delta_{21})$ and $I_3(\Delta_{21})$ since their integrands are obtained by the maximum entropy procedure, although the value of $I_2(\Delta_{21})$ usually remains unknown, since the value $\chi^{(3)}_{\mathrm{NR}}$ is not generally known and cannot be be directly obtained from CARS measurements. However, it follows from (6.62) that

$$\begin{aligned} I_4(\Delta_{21}) &= \int_{\Delta_1}^{\Delta_2} \left[\mathrm{Re}\left\{\chi^{(3)}(\Delta)\right\} - I_1(\Delta_{21})\right]^2 \mathrm{d}\Delta \\ &= \int_{\Delta_1}^{\Delta_2} \left[\mathrm{Re}\left\{\chi^{(3)}(\Delta)\right\} - \chi^{(3)}_{\mathrm{NR}}\right] \mathrm{d}\Delta \end{aligned}$$

$$\mp 2\,|\delta_{21}| \int_{\Delta_1}^{\Delta_2} \left[\mathrm{Re}\left\{\chi^{(3)}(\Delta)\right\} - \chi_{\mathrm{NR}}^{(3)}\right] \mathrm{d}\Delta + |\delta_{21}|^2 \int_{\Delta_1}^{\Delta_2} \mathrm{d}\Delta$$

$$= I_2(\Delta_{21}) - |\delta_{21}|^2 \Delta_{21} \tag{6.64}$$

and, further, from (6.63) and (6.64) that

$$I_2(\Delta_{21}) = I_4(\Delta_{21}) + |\delta_{21}|^2 \Delta_{21} \le I_3(\Delta_{21}) \tag{6.65}$$

or

$$|\delta_{21}|^2 \le \frac{1}{\Delta_{21}} \left[I_3(\Delta_{21}) - I_4(\Delta_{21})\right] = |\Lambda_{21}|^2 . \tag{6.66}$$

Thus, by using the inequality $-|\Lambda_{21}| \le \pm\, |\delta_{21}| \le |\Lambda_{21}|$ and (6.62), the upper and lower limits of the nonresonant susceptibility can be estimated as

$$I_1(\Delta_{21}) - |\Lambda_{21}| \le \chi_{\mathrm{NR}}^{(3)} \le I_1(\Delta_{21}) + |\Lambda_{21}| . \tag{6.67}$$

The validity of this method has been proven in the case of the CARS spectrum of the nitrogen Q-branch [122], as illustrated in Fig. 6.2.

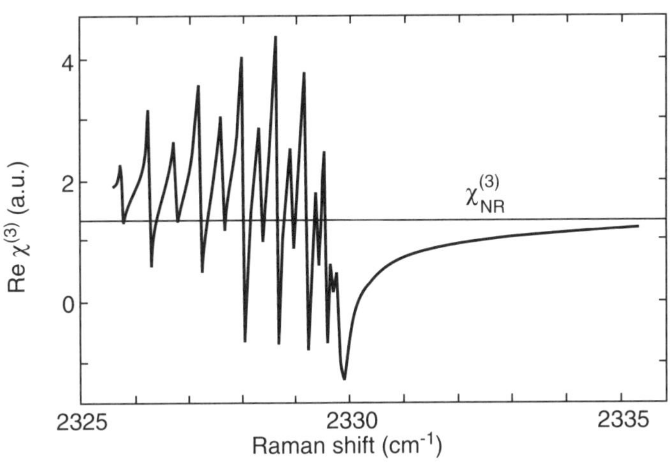

Fig. 6.2. Maximum-entropy estimate of Re $\left\{\chi^{(3)}\right\}$ for nitrogen Q-branch and estimate of $\chi_{\mathrm{NR}}^{(3)}$ obtained by finite-range sum rule analysis

6.6 Poles and Zeros of Meromorphic Susceptibility

In Sect. 3.5 we studied an example of the total susceptibility of a two-level atom. We found out that such a susceptibility can have poles simultaneously in both half planes and in addition a complex zero. It is usually a problem to determine from the measured spectra whether or not there are any complex zeros. From the appearance of the measured intensity spectra, we cannot judge directly the existence of complex zeros. Fortunately, there is the so-called Jensen's formula in complex analysis [13], which helps us to solve this problem. We start by using the conformal mapping of (4.4) and map the upper half plane into the interior of a unit disk. We are forced to consider the meromorphic permittivity of the material. The reasons for this become obvious below. Let us assume that we are dealing with the total meromorphic susceptibility, which contains information about the linear and nonlinear susceptibilities. Then the permittivity is given by the well-known relation $\varepsilon = 1 + \chi$. The total susceptibility can have zeros $\widehat{\omega}_{\text{zero}}$ and poles $\widehat{\omega}_{\text{pole}}$ inside the unit disk. Note carefully that the zeros will be at different locations for ε and χ. If we denote the multiplicity of the poles by n_j and the multiplicity of the zeros by m_k, then we can write Jensen's formula for the meromorphic permittivity as follows [107]:

$$\frac{1}{2\pi}\int_0^{2\pi} \ln\left|\varepsilon\left(\mathrm{e}^{\mathrm{i}\varphi}\right)\right| \mathrm{d}\varphi = \ln|\varepsilon(0)| + \sum_k m_k \ln|\widehat{\omega}_{\text{zero}}|^{-1} - \sum_j n_j \ln|\widehat{\omega}_{\text{pole}}|^{-1}, \tag{6.68}$$

where $W = \mathrm{e}^{\mathrm{i}\varphi}$, $|\widehat{\omega}_{\text{zero}}| < 1$ and $|\widehat{\omega}_{\text{pole}}| < 1$. Furthermore, $\varepsilon(0)$ is neither zero nor infinite. The integrand of (6.68) involves data that are obtained by measurement. At first the estimation of the value of $\varepsilon(0)$ seems to be problematic, since the permittivity calculated at the centre of the unit disk corresponds to the imaginary frequency i in the original angular frequency plane. Fortunately, the purely imaginary frequency matches quite well the zero-angular frequency value of the permittivity in the real physical system. The static value is, practically speaking, governed by the linear susceptibility of the material. Therefore it is possible to estimate the permittivity at the center of the unit disk using a sum rule [40]:

$$\varepsilon(0) = 1 + \frac{2}{\pi}\int_0^{\infty} \frac{\omega \operatorname{Im}\left\{\chi^{(1)}(\omega)\right\}}{\omega^2+1}\mathrm{d}\omega. \tag{6.69}$$

Another method to find the number of poles and zeros relies on the argument theorem [12], which, however, requires information about the complex derivative of the total susceptibility. This is not usually an easy task, because the measured data are obtained only for a real angular frequency variable.

Appendices

A Cauchy's Integral Theorem

If a single-valued holomorphic function $f(z) = u(x,y) + \mathrm{i}v(x,y)$ is given in a singly connected domain D, then the integral of this function along any closed contour C lying totally inside D is equal to zero.

This theorem establishes one of the most important properties of a holomorphic function of a complex variable. Since the function is holomorphic everywhere inside the contour C, it follows that the function possesses continuous first-order partial derivatives in the domain bounded by the contour C. Therefore it is possible to write

$$\begin{aligned}\oint_C f(z)\,\mathrm{d}z &= \oint_C (u\mathrm{d}x - v\mathrm{d}y) + \mathrm{i}(v\mathrm{d}x + u\mathrm{d}y) \\ &= \iint_D \left[\left(-\frac{\partial v}{\partial x} - \frac{\partial u}{\partial y}\right) + \mathrm{i}\left(\frac{\partial u}{\partial x} - \frac{\partial v}{\partial y}\right)\right] \mathrm{d}x\mathrm{d}y = 0\,, \end{aligned} \qquad \text{(A.1)}$$

where the Stokes theorem and Cauchy–Riemann equations have been employed. Actually, the assumption of continuous derivatives is not necessary. Indeed, a more general proof can be given on the basis of the Cauchy–Goursat proof [12].

B Theorem of Residues

If a function f is holomorphic everywhere in a singly connected domain D, except at a finite number of isolated singularities z_k (which are called poles) lying inside the domain, then

$$\oint_C f(z)\,\mathrm{d}z = 2\pi\mathrm{i}\sum_k \operatorname*{Res}_{\text{poles}} [f(z)]\ , \qquad \text{(B.1)}$$

where

$$\operatorname*{Res}_{\text{pole}} f(z) = \frac{1}{(m-1)} \lim_{z \to z_k} \frac{d^{m-1}}{dz^{m-1}} \left[(z - z_k)^m f(z)\right] \qquad \text{(B.2)}$$

is called the residue of the holomorphic function. In (B.2), m denotes the order of the pole. The residue is usually a complex number.

In the event that a single pole is on the integration contour C, the value of the integral (B.1) is equal to $\mathrm{i}\pi \operatorname*{Res}_{\text{pole}} f(z)$.

The key points in the derivation of the above results are the Laurent series expansion and Cauchy's integral formula for the function f. The proofs are found in standard textbooks on complex analysis, e.g. [12, 13].

C Jordan's Lemma

Let k be greater than 0 and consider an integral along the arc of a semicircle in the upper half plane. If it is true that

$$|f(z)| \leq M(B), \ |z| = B, \ \operatorname{Im} z \geq 0,$$

where

$$\lim_{B\to\infty} M(B) = 0 \tag{C.1}$$

and

$$M(B) = \max_{|z|=B, \ \operatorname{Im} z \geq 0} |f(z)| \ , \tag{C.2}$$

then

$$\lim_{B\to\infty} \int_{A_B} f(z)\, \mathrm{e}^{\mathrm{i}kz} \mathrm{d}z = 0, \tag{C.3}$$

where A_B is the arc of the semicircle.

It is possible to estimate that

$$\begin{aligned} |K(B)| &= \left| \int_{A_B} f(z)\, \mathrm{e}^{\mathrm{i}kz} \mathrm{d}z \right| \\ &\leq B \int_0^\pi |f(z)|\, \mathrm{e}^{-kB\sin\varphi} \mathrm{d}\varphi \\ &\leq BM(B) \int_0^\pi \mathrm{e}^{-kB\sin\varphi} \mathrm{d}\varphi. \end{aligned} \tag{C.4}$$

It follows from symmetry and because $(2/\pi)\varphi \leq \sin\varphi$, $0 \leq \varphi \leq \pi/2$, that

$$|K(B)| \leq 2BM(B) \int_0^{\pi/2} \mathrm{e}^{-2kB\varphi/\pi} \mathrm{d}\varphi$$

$$\leq \frac{M(B)\pi}{k}\left(1 - \mathrm{e}^{-kB}\right)$$

$$\leq \frac{\pi}{k} M(B) \to 0, \text{ as } B \to \infty. \tag{C.5}$$

D Phase Retrieval for Oblique-Angle Reflectance: s-Polarization

The electric-field reflectance of s-polarized light is obtained from the equation

$$r_{\mathrm{s}} = \frac{\cos\theta - \sqrt{N^2 - \sin^2\theta}}{\cos\theta + \sqrt{N^2 - \sin^2\theta}}. \tag{D.1}$$

We can set

$$\sqrt{N^2 - \sin^2\theta} = a + \mathrm{i}b. \tag{D.2}$$

By measurement we get

$$R_{\mathrm{s}} = r_{\mathrm{s}} r_{\mathrm{s}}^{*} = \frac{(a - \cos\theta)^2 + b^2}{(a + \cos\theta)^2 + b^2}. \tag{D.3}$$

Then we can write that

$$r_{\mathrm{s}} = |r_{\mathrm{s}}|\,\mathrm{e}^{\mathrm{i}\varphi} = |R_{\mathrm{s}}|^{1/2}\,\mathrm{e}^{\mathrm{i}\varphi} = c + \mathrm{i}d = \frac{\cos\varphi - (a + \mathrm{i}b)}{\cos\varphi + (a + \mathrm{i}b)}. \tag{D.4}$$

Resolving the real and imaginary parts, we find that

$$a = \frac{1 - e}{1 + e}\cos\theta\,,$$
$$b = \frac{-2d}{(1 + c)(1 + e)}\cos\theta, \tag{D.5}$$

where

$$e = c + \frac{\mathrm{d}^2}{1 + c}. \tag{D.6}$$

By using the formulae (D.2), (D.4), (D.5) and (D.6) the complex refractive index can be calculated in a phase retrieval procedure, after φ has been obtained.

E Complex Analysis with Several Variables

Complex analysis with several variables is a natural extension of the theory of one complex variable. In the case of several complex variables we deal with the n-dimensional complex space C^n, where apoint, $z = (z_1, z_2, ..., z_n)$ is defined as $z_k = x_k + iy_k$. Therefore the complex space can be considered as an ordinary Euclidean space of dimension $2n$. A function $f(z) = u(x_1, y_1, ..., x_n, y_n) + iv(x_1, y_1, ..., x_n, y_n)$ of several complex variables is defined in such a way that every point of the n-dimensional space is associated with a definite complex number that belongs to C^1.

A holomorphic function of several complex variables is defined in a similar manner to the case of one complex variable. A function f of several complex variables is holomorphic provided that the Cauchy–Riemann conditions hold for all variables as follows:

$$\frac{\partial u}{\partial x_k} = \frac{\partial v}{\partial y_k},$$
$$\frac{\partial u}{\partial y_k} = -\frac{\partial v}{\partial x_k}. \tag{E.1}$$

Cauchy's integral formula can be generalized to holomorphic functions of several variables. As an example, suppose that $f(z_1, z_2)$ is holomorphic in a domain $D = D_1 \times D_2$, where D_1 and D_2 are simply connected domains. We take arbitrary closed contours C_1 and C_2 in D_1 and D_2 and consider the integral

$$I = \oint_{C_1} \mathrm{d}z_1 \oint_{C_2} \frac{f(z_1, z_2)}{(z_1 - z_1')(z_2 - z_2')} \mathrm{d}z_2, \tag{E.2}$$

where z_1' and z_1' are arbitrary points lying inside the contours. Since the function f is holomorphic, the inner integral in (E.2) can be expressed, with the aid of Cauchy's formula for one complex variable, by

$$\oint_{C_2} \frac{f(z_1, z_2)}{(z_1 - z_1')(z_2 - z_2')} \mathrm{d}z_2 = 2\pi \mathrm{i} \frac{f(z_1, z_2')}{z_1 - z_1'}. \tag{E.3}$$

Making use of Cauchy's formula once again, we finally get

$$I = (-4\pi^2) f(z_1' z_2'). \tag{E.4}$$

Similarly, in the case of n variables, we have the formula

$$f(z_1', ..., z_n') = \frac{1}{(2\pi \mathrm{i})^n} \oint_{C_1} \mathrm{d}z_1 ... \oint_{C_n} \frac{f(z_1, ..., z_n)\,\mathrm{d}z_n}{\prod_{k=1}^{n} (z_k - z_k')}, \tag{E.5}$$

which is the generalization of Cauchy's formula.

A holomorphic function of several complex variables has partial derivatives of all orders. In the case of two variables the following expression is then true:

$$\left.\frac{\partial^{j+l} f(z_1,\ z_2)}{\partial z_1^j \partial z_2^l}\right|_{(z_1',z_2')} = -\frac{j!\ l!}{4\pi^2} \oint_{C_1} \mathrm{d}z_1 \oint_{C_2} \frac{f(z_1,\ z_2)\,\mathrm{d}z_2}{(z_1 - z_1')^{j+1}(z_2 - z_2')^{l+1}}. \tag{E.6}$$

Series expansions for holomorphic functions of several complex variables can be constructed. The Taylor series of the function f of two complex variables in a polydisk is given by

$$f(z_1,\ z_2) = \sum_{j=0}^{\infty} \sum_{l=0}^{\infty} c_{jl} z_1^j z_2^l, \tag{E.7}$$

where, according to (E6),

$$c_{jl} = \frac{1}{j!\ l!} \left.\frac{\partial^{j+l} f(z_1,\ z_2)}{\partial z_1^j \partial z_2^l}\right|_{(0,0)} \tag{E.8}$$

when the series expansion is developed around $(0, 0)$.

F Surface Integral of Generalized Cauchy Formula

We consider the simple case of a surface integral in (4.35) involving a linear term W_1 only. Then the complex derivative of (4.36) is equal to one. It is helpful to consider the situation shown in Fig. F.1.

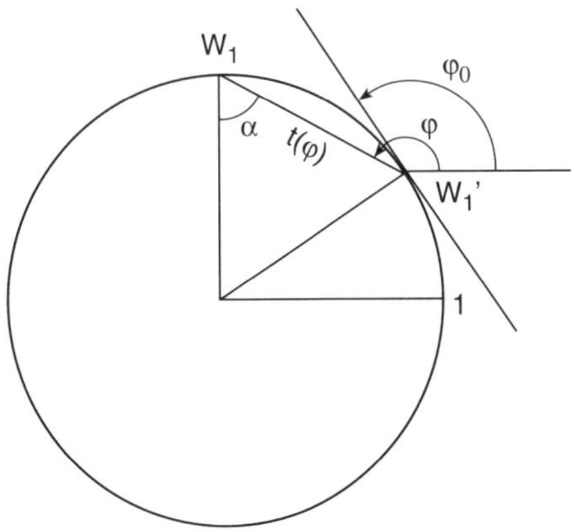

Fig. F.1. Calculation of surface integral

We set

$$w_1 = u_1 + \mathrm{i}v_1 = W_1 - W_1' = r\mathrm{e}^{\mathrm{i}\varphi} \tag{F.1}$$

and then the surface integral can be written in the following way:

$$I = -\frac{1}{\pi} \iint_{|W_1|\leq 1} \frac{\mathrm{d}u_1 \mathrm{d}v_2}{w_1} \,. \tag{F.2}$$

On the other hand, we find from Fig. F.1 that

$$\begin{aligned} \varphi_0 &= \arg\{W_1'\} + \pi/2 \,, \\ \alpha &= -\pi + (\varphi - \arg\{W_1'\}) = \varphi - \arg\{-W_1'\} \,, \\ t(\varphi) &= 2r\cos\alpha = 2r\cos(\varphi - \arg\{-W_1'\}) \,. \end{aligned} \tag{F.3}$$

Then we have also

$$I = -\frac{1}{\pi} \int\limits_{\varphi_0}^{\varphi_0+\pi} \int\limits_{\varphi_0}^{t(\varphi)} \frac{r\mathrm{d}r\mathrm{d}\varphi}{r\mathrm{e}^{\mathrm{i}\varphi}} = -\frac{1}{\pi} \int\limits_{\varphi_0}^{\varphi_0+\pi} \frac{t(\varphi)}{\mathrm{e}^{\mathrm{i}\varphi}} \mathrm{d}\varphi \,. \tag{F.4}$$

We next substitute $\varphi = \alpha + \arg\{-W_1'\}$ from (F.3) into (F.4), and then we obtain

$$\begin{aligned} I &= -\frac{2r}{\pi} \int\limits_{-\pi/2}^{\pi/2} \mathrm{e}^{\mathrm{i}(-\alpha - \arg\{-W_1'\})} \cos\alpha \mathrm{d}\alpha \\ &= -\frac{2r}{\pi} \mathrm{e}^{-\mathrm{i}\arg\{-W_1'\}} \int\limits_{-\pi/2}^{\pi/2} \mathrm{e}^{-\mathrm{i}\alpha} \cos\alpha \mathrm{d}\alpha . \end{aligned} \tag{F.5}$$

Since

$$r\mathrm{e}^{-\mathrm{i}\arg\{-W_1'\}} = \frac{W_1' W_1'^*}{|-W_1'|\,\mathrm{e}^{\mathrm{i}\arg\{-W_1'\}}} = \frac{W_1' W_1'^*}{-W_1'} = -W_1'^* \tag{F.6}$$

and

$$\int\limits_{-\pi/2}^{\pi/2} \left(\cos^2\alpha - \mathrm{i}\sin\alpha\cos\alpha\right) \mathrm{d}\alpha = \pi/2, \tag{F.7}$$

we find the required result $I = W_1'^*$.

G Derivation of (4.40)

$$
\begin{aligned}
\left|\chi_{\mathrm{R}}^{(3)}(\varphi)\right|^2 &= \chi_{\mathrm{R}}^{(3)}(\varphi)\,\chi_{\mathrm{R}}^{(3)}(\varphi)^* = \left(\sum_{j=0}^{J} c_j \mathrm{e}^{\mathrm{i}j\varphi}\right)\left(\sum_{j=0}^{J} c_j{}^* \mathrm{e}^{-\mathrm{i}j\varphi}\right) \\
&= c_0 \sum c_j{}^* \mathrm{e}^{-\mathrm{i}j\varphi} + c_1 \varphi^{\mathrm{i}\varphi} \sum c_j{}^* \mathrm{e}^{-\mathrm{i}j\varphi} + \ldots \\
&\quad + c_J \mathrm{e}^{\mathrm{i}J\varphi} \sum c_j{}^* \mathrm{e}^{-\mathrm{i}j\varphi} \\
&= c_0 c_0^* + c_0 c_1^* \mathrm{e}^{-\mathrm{i}\varphi} + \ldots + c_0 c_J^* \mathrm{e}^{-\mathrm{i}J\varphi} \\
&\quad + c_1 c_0^* \mathrm{e}^{\mathrm{i}\varphi} + c_1 c_1^* + \ldots + c_1 c_J^* \mathrm{e}^{-\mathrm{i}(J-1)\varphi} + \ldots \\
&\quad + c_l c_0^* \mathrm{e}^{\mathrm{i}l\varphi} + c_l c_1^* \mathrm{e}^{\mathrm{i}(l-1)\varphi} + \ldots + c_l c_l^* + \ldots \\
&\quad + c_l c_J{}^* \mathrm{e}^{-\mathrm{i}(J-l)\varphi} + \ldots \\
&\quad + c_J c_0^* \mathrm{e}^{\mathrm{i}J\varphi} + c_J c_1^* \mathrm{e}^{\mathrm{i}(J-1)\varphi} + \ldots + c_J c_J^* \\
&= c_0 c_0^* + c_1 c_1^* + \ldots + c_J c_J^* \\
&\quad + \left(c_0 c_1^* + \ldots + c_{J-1} c_J^*\right) \mathrm{e}^{-\mathrm{i}\varphi} + \left(c_1 c_0^* + \ldots\right. \\
&\quad \left. + \, c_J c_{J-1}^*\right) \mathrm{e}^{\mathrm{i}\varphi} + \ldots \\
&\quad + \left(c_0 c_l^* + \ldots + c_{J-l} c_J^*\right) \mathrm{e}^{-\mathrm{i}l\varphi} + \left(c_l c_0^* + \ldots\right. \\
&\quad \left. + \, c_J c_{J-l}^*\right) \mathrm{e}^{\mathrm{i}l\varphi} + \ldots \\
&\quad + c_0 c_J^* \mathrm{e}^{-\mathrm{i}J\varphi} + c_J c_0^* \mathrm{e}^{\mathrm{i}J\varphi}.
\end{aligned}
\tag{G.1}
$$

Substituting $c_l = a_l + \mathrm{i}b_l$ and $c_l{}^* = a_l - \mathrm{i}b_l$ we obtain

$$
\begin{aligned}
\left|\chi_{\mathrm{R}}^{(3)}(\varphi)\right|^2 &= a_0^2 + b_0^2 + \ldots + a_J^2 + b_J^2 \\
&\quad + [a_0 a_1 + b_0 b_1 + \ldots + a_{J-1} a_J + b_{J-1} b_J
\end{aligned}
$$

$$+\mathrm{i}\left(a_1 b_0 - a_0 b_1 + \ldots + a_J b_{J-1} - a_{J-1} b_J\right)\right] \mathrm{e}^{-\mathrm{i}\varphi}$$

$$+\left[a_0 a_1 + b_0 b_1 + \ldots + a_{J-1} a_J + b_{J-1} b_J\right.$$

$$-\mathrm{i}\left(a_1 b_0 - a_0 b_1 + \ldots + a_J b_{J-1} - a_{J-1} b_J\right)\right] \mathrm{e}^{\mathrm{i}\varphi}$$

$$+\ldots + \left[a_0 a_J + b_0 b_J + \mathrm{i}\left(a_J b_0 - a_0 b_J\right)\right] \mathrm{e}^{-\mathrm{i}J\varphi}$$

$$+\left[a_0 a_J + b_0 b_J - \mathrm{i}\left(a_J b_0 - a_0 b_J\right)\right] \mathrm{e}^{\mathrm{i}J\varphi}. \tag{G.2}$$

If we now set

$$A'_j = \sum_{l=0}^{J-j} \left(a_l a_{l+j} + b_l b_{l+j}\right), \quad j = 0,\ 1,\ 2, \ldots, J \tag{G.3}$$

and

$$B'_j = \sum_{l=0}^{J-j} \left(a_{l+j} b_l - a_l b_{l+j}\right), \quad j = 1,\ 2, \ldots, J, \tag{G.4}$$

the expression for $\left|\chi_{\mathrm{R}}^{(3)}(\varphi)\right|^2$ is simplified to the form

$$\begin{aligned}\left|\chi_{\mathrm{R}}^{(3)}(\varphi)\right|^2 &= A'_0 + \left(A'_1 + \mathrm{i}B'_1\right)\mathrm{e}^{-\mathrm{i}\varphi} + \left(A'_1 - \mathrm{i}B'_1\right)\mathrm{e}^{\mathrm{i}\varphi} + \ldots \\ &\quad + \left(A'_J + \mathrm{i}B'_J\right)\mathrm{e}^{-\mathrm{i}J\varphi} + \left(A'_J - \mathrm{i}B'_J\right)\mathrm{e}^{\mathrm{i}J\varphi} \\ &= A'_0 + A'_1\left(\mathrm{e}^{\mathrm{i}\varphi} + \mathrm{e}^{-\mathrm{i}\varphi}\right) + \mathrm{i}B'_1\left(\mathrm{e}^{-\mathrm{i}\varphi} - \mathrm{e}^{\mathrm{i}\varphi}\right) + \ldots \\ &\quad + A'_J\left(\mathrm{e}^{\mathrm{i}J\varphi} + \mathrm{e}^{-\mathrm{i}J\varphi}\right) + \mathrm{i}B'_J\left(\mathrm{e}^{-\mathrm{i}J\varphi} - \mathrm{e}^{\mathrm{i}J\varphi}\right).\end{aligned} \tag{G.5}$$

Using the Euler equations $\mathrm{e}^{-\mathrm{i}\varphi} - \mathrm{e}^{\mathrm{i}\varphi} = -2\mathrm{i}\sin\varphi$ and $\mathrm{e}^{\mathrm{i}\varphi} + \mathrm{e}^{-\mathrm{i}\varphi} = 2\cos\varphi$, we finally obtain (4.40).

H Derivation of (5.21)

$$\begin{aligned}\left|A_M(z)\right|^2 &= A_M(z)\, A_M^*(1/z^*) = \left(\sum_{n=0}^{M} \alpha_n z^{-n}\right)\left(\sum_{n=0}^{M} \alpha_n^* z^n\right) \\ &= \alpha_0 \sum a_n^* z^n + \alpha_1 z^{-1} \sum a_n^* z^n + \cdots + \alpha_M z^{-M} \sum a_n^* z^n\end{aligned}$$

$$
\begin{aligned}
&= \alpha_0\alpha_0^* + \alpha_0\alpha_1^* z + \cdots + \alpha_0 c_M^* z^M \\
&\quad + \alpha_1\alpha_0^* z^{-1} + \alpha_1\alpha_1^* + \cdots + \alpha_1\alpha_M^* z^{M-1} \\
&\quad + \cdots \\
&\quad + \alpha_k\alpha_0^* z^{-k} + \alpha_k\alpha_1^* z^{1-k} + \cdots + \alpha_k\alpha_k^* + \cdots + \alpha_k\alpha_M^* z^{M-k} \\
&\quad + \cdots \\
&\quad + \alpha_M\alpha_0^* z^{-M} + \alpha_N\alpha_1^* z^{1-M} + \cdots + \alpha_M\alpha_M^* \\
&= \alpha_0\alpha_0^* + \alpha_1\alpha_1^* + \cdots + \alpha_M\alpha_M^* \\
&\quad + \left(\alpha_0^*\alpha_1 + \cdots + \alpha_{M-1}^*\alpha_M\right) z^{-1} \\
&\quad + \left(\alpha_0\alpha_1^* + \cdots + \alpha_{M-1}\alpha_M^*\right) z \\
&\quad + \cdots \\
&\quad + \left(\alpha_0^*\alpha_k + \cdots + \alpha_{M-k}^*\alpha_M\right) z^{-k} \\
&\quad + \left(\alpha_0\alpha_k^* + \cdots + \alpha_{M-k}\alpha_M^*\right) z^k \\
&\quad + \cdots \\
&\quad + \alpha_0^*\alpha_M z^{-M} + \alpha_0\alpha_M^* z^M .
\end{aligned}
\tag{H.1}
$$

If we now set

$$
a_n = \sum_{k=0}^{M-n} \alpha_k^* \alpha_{k+n}, \quad 0 \le n \le M \ , \tag{H.2}
$$

we see that

$$
a_{-n} = \sum_{k=0}^{M-n} \alpha_k \alpha_{k+n}^* = a_n^* . \tag{H.3}
$$

Hence, we obtain

$$|A_M(z)|^2 = A_M(z)\,A_M^*(1/z^*) = \sum_{n=-M}^{M} a_n z^n. \tag{H.4}$$

I Superconvergence Theorem [33]

If g is a function defined by

$$g(x) = \mathrm{P}\int_0^\infty \frac{f(y)}{y^2 - x^2}\,\mathrm{d}y \tag{I.1}$$

and $f(y)$ falls off at large y faster than $(y \ln y)^{-1}$, then for large x, the following expansion is valid:

$$g(x) = -\frac{1}{x^2}\int_0^\infty f(y)\,\mathrm{d}y + O\left(x^{-2}\right)\,, \tag{I.2}$$

where the remainder term O denotes terms falling off faster than x^{-2}.

References

1. Wooten F. (1972) Optical Properties of Solids, Academic Press, New York
2. Kittel C. (1976) Introduction to Solid State Physics, 5th edn, Wiley, New York
3. Smith D. Y., Dexter. D. L. (1972) Optical and Absorption Strength of Defects in Insulators. In: E. Wolf (Ed) Progress in Optics, Vol X, North-Holland, Amsterdam
4. Robinson P. W., Reid G. T. (Eds) (1993) Interferogram Analysis, IOP Publishing, Bristol
5. Asakura T. (1976) Surface Roughness Measurement. In: R. K. Erf (Ed) Speckle Metrology, Academic Press, Orlando
6. Räsänen J., Peiponen K.-E., Silvennoinen R., Silfsten P. (1992) Crystal Surface Quality Inspection Using Computer-Generated Hologram and Digital Image Processing, Phys Stat Sol (a) 134:K53
7. Sommerfeld A. (1914) Über die Fortpflanzung des Lichtes in dispergierenden Medien, Ann Phys 44:177
8. Brillouin L. (1914) Über die Fortpflanzung des Lichtes in dispergierenden Medien, Ann Phys 44:203
9. Arfken G. (1970) Mathematical Methods for the Physicists, 2nd edn, Academic Press, New York
10. Nussenzveig H. M. (1972) Causality and Dispersion Relations, Academic Press, New York
11. Loudon R. (1973) The Quantum Theory of Light, Clarendon Press, Oxford
12. Ahlfors L. (1979) Complex Analysis, 3rd edn, McGraw-Hill, New York
13. Nevanlinna R. (1970) Analytic Functions, Springer, Berlin, Heidelberg
14. Born M., Wolf E. (1980) Principles of Optics, 6th edn, Pergamon, New York
15. Holm R. T. (1998) Conventions Confusions. In: E. W. Palik (Ed) Handbook of Optical Constants of Solids II, Academic Press, Orlando
16. Azzam R. M. A. (1979) Direct Relation between Fresnel's Interface Reflection Coefficients for the Parallel and Perpendicular Polarizations, J Opt Soc Am 69:1007
17. Humphreys-Owen S. P. F. (1961) Comparison of Reflection Methods for Measuring Optical Constants without Polarimetric Analysis, and Proposal for New Methods Based on the Brewster Angle, Proc Phys Soc 77:949
18. Beckmann P., Spizzichino A. (1963) The Scattering of Electromagnetic Waves from Rough Surfaces, Pergamon, Oxford
19. Cielo P. (1988) Optical Techniques for Industrial Inspection, Academic Press, San Diego
20. Kronig R. (1926) On the Theory of Dispersion of X-Rays, J Opt Soc Am 12:547
21. Kramers H. A. (1929) Die Dispersion und Absorption von Röntgenstralen, Phys Z 30:522

22. Kronig R. (1942) Algemeene Theorie der Dielecrische en Magnetische Verliezen, Ned T Natuurk 9:402
23. Titchmarsh E. C. (1948) Introduction to the Theory of Fourier Integrals, 2nd edn, Clarendon Press, Oxford
24. Toll J. S. (1956) Causality and Dispersion Relation: Logical Foundations, Phys Rev 104:1760
25. Narashimhan R. (1985) Complex Analysis in One Variable, Birkhäuser, Boston
26. Remacle F., Levine R. D. (1993) Time Domain Information from Resonant Raman Excitation Profiles: A Direct Inversion by Maximum Entropy, J Chem Phys 99:4908
27. Lee S.-Y. (1995) Phase Recovery and Reconstruction of the Raman Amplitude from Raman Excitation Profile, Chem Phys Lett 245:620
28. Kircheva P. P., Hadjichristov G. B. (1994) Kramers–Kronig Relations in FWM Spectroscopy, J Phys B: Mol Opt Phys 27:3781
29. Morse P. M., Feshbach H. (1953) Methods of Theoretical Physics, McGraw-Hill, New York
30. Mandel L., Wolf E. (1965) Coherence Properties of Optical Fields, Rev Mod Phys 37:231
31. Klauder J. R., Sudarshan E. C. G. (1968) Fundamentals of Quantum Optics, W.A. Benjamin, New York
32. Bohren C. F., Huffman D. R. (1983) Absorption and Scattering of Light by Small Particles, Wiley, New York
33. Bassani F., Altarelli M. (1983) Interaction of Radiation with Condensed Matter. In: E. E. Koch (Ed) Handbook on Synchrotron Radiation, Vol 1(a), North-Holland, Amsterdam
34. Peiponen K.-E., Vartiainen E. M. (1991) Kramers–Kronig Relations in Optical Data Inversion, Phys Rev B44:8301
35. Lüty F. (1968) F_A Centres in Alkali Halide Crystals. In: W. B. Fowler (Ed) Physics of Color Centers, Academic Press, New York
36. Peiponen K.-E., Silfsten P., Luostarinen J., Vartiainen E. M., Ketolainen P., Asakura T. (1992) Ruby Laser-Induced Frequency Dependent Refractive Index Change in KCl:Li+ F_A Crystal, Phys Stat Sol (b) 174:K39
37. Ahrenkiel R. K. (1971) Modified Kramers–Kronig Analysis of Optical Spectra, J Opt Soc Am 61:1651
38. Goplen T. G., Cameron D. G., Jones R. N. (1980) The Control of Errors in Infrared Spectrophotometry. VI. The Evaluation of Optical Constants by Combined Transmission and Attenuated Total Reflection Measurements, Appl Spectr 34:653
39. Hulthen R. (1982) Kramers–Kronig Relations Generalized: on Dispersion Relations for Finite Frequency Intervals. A Spectrum-Restoring Filter, J Opt Soc Am 72:794
40. King F. W. (1976) Sum Rules for Optical Constants, J Math Phys 17:1509
41. Räty J., Keränen E., Peiponen K.-E. (1998) The Complex Refractive Index Measurement of Liquids by a Novel Reflectometer Apparatus for the UV–Visible Spectral Range, Meas Sci Technol 9:95
42. Velický B. (1961) Dispersion Relation for Complex Reflectivity, Czech J Phys B11:541
43. Smith D. Y. (1977) Dispersion Relations for Complex Reflectivities, J Opt Soc Am 67:570
44. Roessler D. M. (1965) Kramers–Kronig Analysis of Reflection Data, Brit J Appl Phys 16:1119

45. Zeng X. C., Bergman D. J, Hui P. M., Stroud D. (1988) Effective-Medium Theory for Weakly Nonlinear Composites, Phys Rev B38:10970
46. Haus J. W., Inguva R., Bowden C. M. (1989) Effective-Medium Theory of Nonlinear Ellipsoidal Composites, Phys Rev A40:5729
47. Sipe J. E., Boyd R. W. (1992) Nonlinear Susceptibility of Composite Optical Materials in the Maxwell Garnett Model, Phys Rev A46:1614
48. Fischer G. L., Boyd R. W., Gehr R. J., Jenehke S. A., Osaheni J. A., Sipe J. E., Weller-Brophy, L. A. (1995) Enhanced Nonlinear Optical Response of Composite Materials, Phys Rev Lett 74:1871
49. Gehr R. J., Fischer G. L., Boyd R. W. (1997) Nonlinear-Optical Response of Porous-Glass-Based Composite Materials, J Opt Soc Am B14:2310
50. Smith D. D., Fischer G., Boyd R. W., Gregory D. A. (1997) Cancellation of Photoinduced Absorption in Metal Nanoparticle Composites through a Counterintuitive Consequence of Local Field Effects, J Opt Soc Am B14:1625
51. Maxwell-Garnett J. C. (1904) Colours in Metal Glasses and in Metallic Films, Trans R Soc 203:385
52. Maxwell-Garnett J. C. (1906) Colours in Metallic Glasses, in Metallic Films, and in Metallic Solution, Trans R Soc 205:237
53. Smith D. Y. (1976) Comments on the Dispersion Relations for the Complex Refractive Index of Circularly and Elliptically Polarized Light, J Opt Soc Am 66:454
54. Smith D. Y. (1976) Dispersion Relations for Magnetoreflectivity, J Opt Soc Am 66:547
55. Shen Y. R. (1984) The Principles of Nonlinear Optics, Wiley, New York
56. Lamb G. L., Jr. (1980) Elements of Soliton Theory, Wiley, New York
57. Krantz S. G. (1982) Function Theory of Several Complex Variables, Wiley, New York
58. Kogan M. (1963) On the Electrodynamics of Weakly Nonlinear Media, Sov Phys JETP 16:217
59. Price P. J. (1963) Theory of Quadratic Response Functions, Phys Rev 130:1792
60. Caspers W. J. (1964) Dispersion Relations for Nonlinear Response, Phys Rev 133:A1249
61. Ridener F. L., Good R. H., Jr. (1975) Dispersion Relations for Nonlinear Systems of Arbitrary Degree, Phys Rev B11:2768
62. Smet F., van Groenendael A. (1979) Dispersion Relations for n-Order Nonlinear Phenomena, Phys Rev A19:334
63. Peiponen K.-E. (1987) Sum Rules for the Nonlinear Susceptibilities in the Case of Sum Frequency Generation, Phys Rev B35:4116
64. Peiponen K.-E. (1988) Nonlinear Susceptibilities as a Function of Several Complex Angular Frequency Variables, Phys Rev B37:6463
65. Peiponen K.-E., Vartiainen E. M., Asakura T. (1997) Complex Analysis in Dispersion Theory, Opt Rev 4:433
66. Kishida H., Hasegawa T., Iwasa Y., Koda T. (1993) Dispersion Relation in the Third-Order Electric Susceptibility for Polysilane Film, Phys Rev Lett 70:3724
67. Miller R. D., Michl J. (1989) Polysilane High Polymers, Chem Rev 89:1359
68. Bassani F., Scandolo S. (1991) Dispersion Relations and Sum Rules in Nonlinear Optics, Phys Rev B44:8446
69. Yariv A. (1967) Quantum Electronics, 2nd edn, Wiley, New York
70. Hutchings D. C., Sheik-Bahae M., Hagan D. J., van Stryland E. W. (1992) Kramers–Kronig Relations in Nonlinear Optics, Opt Quant Electron 24:1

71. Shore K. A., Chan D. A. S. (1990) Kramers–Kronig Relations for Non-Linear Optics, Electron Lett 26:1206
72. Tokunaga E., Terasaki A., Kobayashi T. (1993) Femtosecond Time-Resolved Dispersion Relations Studied with a Frequency-Domain Interferometer, Phys Rev A47:R4581
73. Tokunaga E., Terasaki A., Kobayashi T. (1995) Femtosecond Phase Spectroscopy by Use of Frequency-Domain Interference, J Opt Soc Am B12:753
74. Tokunaga E., Terasaki A., Kobayashi T. (1996) Femtosecond Continuum Interferometer for Transient Phase and Transmission Spectroscopy, J Opt Soc Am B13:496
75. Silverman H. (1975) Complex Variables, Houghton Mifflin, Boston
76. Gonzales M. O. (1992) Complex Analysis: Selected Topics, Marcel Dekker, New York
77. Vartiainen E. M., Peiponen K.-E. (1994) Meromorphic Degenerate Nonlinear Susceptibility: Phase Retrieval from the Amplitude Spectrum, Phys Rev B50:1941
78. Peiponen K.-E., Vartiainen E. M., Asakura T. (1997) Dispersion Relations and Phase Retrieval in Optical Spectroscopy. In: E. Wolf (Ed) Progress in Optics XXXVII, Elsevier Science, Amsterdam
79. Hopf F. A., Stegeman G. I. (1986) Applied Classical Electrodynamics, Nonlinear Optics, Wiley, New York
80. King W. F. (1978) Analysis of Optical Data by the Conjugate Fourier-Series Approach, J Opt Soc Am 68:994
81. Ketolainen P., Peiponen K.-E., Karttunen K. (1991) Refractive-Index Change due to F Color Centers in KCl_xBr_{1-x} Mixed Crystals, Phys Rev B43:4492
82. King F. W. (1977) A Fourier Series Algorithm for the Analysis of Reflectance Data, J Phys C: Solid State Phys 10:3199
83. Azzam R. M. A. (1980) Mapping of Fresnel's Interface Reflection Coefficients between Normal and Oblique Incidence: Results for the Parallel and Perpendicular Polarizations at Several Angles of Incidence, Appl Opt 19:3361
84. Peiponen K.-E., Vartiainen E. M., Tsuboi T. (1990) Sum Rules for the Squared Modulus of the Nonlinear Raman Susceptibility, Phys Rev A41:527
85. Lehto O., Virtanen K. I. (1973) Quasiconformal Mappings in the Plane, Springer, Berlin, Heidelberg
86. Vartiainen E. M., Peiponen K.-E., Tsuboi T. (1990) Analysis of Coherent Raman Spectra, J Opt Soc Am B7:722
87. Vartiainen E. M. (1992) Phase Retrieval Approach for Coherent Anti-Stokes Raman Scattering Spectrum Analysis, J Opt Soc Am B9:1209
88. Vartiainen E. M., Peiponen K.-E., Asakura T. (1992) Maximum Entropy Model in Reflection Spectra Analysis, Opt Commun 89:37
89. Vartiainen E. M., Peiponen K.-E., Asakura T. (1993) Comparison between the Optical Constants Obtained by the Kramers–Kronig Analysis and the Maximum Entropy Method: Infrared Optical Properties of Orthorhombic Sulfur, Appl Opt 32:1126
90. Vartiainen E. M., Asakura T., Peiponen K.-E. (1993) Generalized Noniterative Maximum Entropy Procedure for Phase Retrieval Problems in Optical Spectroscopy, Opt Commun 104:149
91. Shannon C. E. (1948) A Mathematical Theory of Communication, Bell Syst Tech J 27:623
92. Jaynes E. T. (1957) Information Theory and Statistical Mechanics, Phys Rev 106:620
93. Jaynes E. T. (1957) On the Rationale of Maximum-Entropy Methods, Proc IEEE 70:939

94. Gull S. F., Daniel G. J. (1978) Image Reconstruction from Incomplete and Noisy Data, Nature 272:686
95. Bryan R. K., Skilling J. (1986) Maximum Entropy Image Reconstruction From Phaseless Fourier Data, Opt Acta 33:287
96. Collins D. M. (1985) Electron Density Images from Imperfect Data by Iterative Entropy Maximization, Nature 298:49
97. Kemp M. C. (1981) Maximum Entropy Reconstruction in Emission Tomography. In: Medical Radionucleic Imaging Publ. IAEA-SM 247, Vol 1:128
98. Matsumoto T., Tanji T., Tonomura A. (1994) Application of Maximum Entropy Method (MEM) to In-Line Fresnel Electron Holography, Optik 97:169
99. Burg J. P. (1967) Maximum Entropy Spectral Analysis. 37th Ann Int Meeting Soc Explor Geophysics, Oklahoma City, OK
100. Haykin S., Kesler S. (1983) Prediction-Error Filtering and Maximum-Entropy Spectral Estimation. In: S. Haykin (Ed) Nonlinear Methods of Spectral Analysis, 2nd edn, Springer, Berlin, Heidelberg
101. van den Bos A. (1971) Alternative Interpretation of Maximum Entropy Spectral Analysis, IEEE Trans Inf Theory IT-17:493
102. Press W. H., Teukolsky S. A., Vetterling W. T., Flannery B. P. (1989) Numerical Recipes, Cambridge University Press, Cambridge
103. Vartiainen E. M., Peiponen K.-E., Asakura T. (1996) Phase Retrieval in Optical Spectroscopy: Resolving Optical Constants from Power Spectra, Appl Spectr 50:1283
104. Palik E. D. (1985) Gallium Arsenide. In: E. D. Palik (Ed) Handboook of Optical Constants of Solids, Academic Press, London
105. Palik E. D. (1985) Potassium Chloride. In: E. D. Palik (Ed) Handbook of Optical Constants of Solids, Academic Press, London
106. Vartiainen E. M., Peiponen K.-E., Kishida H., Koda T. (1996) Phase Retrieval in Nonlinear Optical Spectroscopy by the Maximum-Entropy Method: an Application to the $\left|\chi^{(3)}\right|$ Spectra of Polysilane, J Opt Soc Am B13:2106
107. Peiponen K.-E., Vartiainen E. M., Asakura T. (1997) Dispersion Theory and Phase Retrieval of Meromorphic Total Susceptibility, J Phys: Condens Matter 9:8937
108. Boyd R. W., Sipe J. E. (1994) Nonlinear Optical Susceptibilities of Layered Composite Materials, J Opt Soc Am B11:297
109. De Alfaro V., Fubini G., Rosetti G., Furlan G. (1966) Sum Rules for Strong Interactions, Phys Lett 21:576
110. Altarelli M., Dexter D. L., Nussenzveig H. M., Smith D. Y. (1972) Superconvergence and Sum Rules for the Optical Constants, Phys Rev B6:4502
111. Altarelli M., Smith D. Y. (1974) Superconvergence and Sum Rules for the Optical Constants: Physical Meaning, Comparison with Experiment, and Generalization, Phys Rev B9:1290
112. Peiponen K.-E (1985) A Simple Derivation of Some Sum Rules for the Powers of the Optical Constants of Insulators, Lett Nuovo Cimento 44:445
113. Villani A., Zimerman A. H. (1973) Generalized f-Sum Rules for the Optical Constants, Phys Lett 44A:295
114. Gründler R. (1983) Superconvergent Sum Rules for Metals and Insulators, Phys Stat Sol (b) 115:K147
115. Peiponen K.-E. (1987) On the Derivation of Sum Rules for Physical Quantities and Their Applications for Linear and Non-Linear Optical Constants, J Phys C: Solid State Phys 20:2785
116. Smith D. Y., Manogue C. A. (1981) Superconvergence Relations and Sum Rules for Reflection Spectroscopy, J Opt Soc Am 71:935

117. King F. W. (1979) Dispersion Relations and Sum Rules for the Normal Reflectance of Conductors and Insulators, J Chem Phys 71:4726
118. Smith D. Y. (1976) Superconvergence and Sum Rules for the Optical Constants: Natural and Magneto-Optical Activity, Phys Rev B13:5303
119. Furuya K., Villani A., Zimerman A. H. (1977) Superconvergent Sum Rules for the Normal Reflectivity, J Phys C: Solid State Phys 10:3189
120. Peiponen K.-E. (1987) Sum Rules for Non-Linear Susceptibilities in the Case of Difference-Frequency Generation, J Phys C: Solid State Phys 20:L285
121. Scandolo S., Bassani F. (1992) Nonlinear Sum Rules: The Three Level and Anharmonic Oscillator Models, Phys Rev B45:13257
122. Vartiainen E. M., Peiponen K.-E., Asakura T. (1993) Sum Rules in Testing Non-Linear Susceptibility Obtained Using Maximum Entropy Model, J Phys: Condens Matter 5:L113

Index

Springer Tracts in Modern Physics

130 **Time-Resolved Light Scattering from Excitons**
By H. Stolz 1994. 87 figs. XI, 210 pages

131 **Ultrathin Metal Films**
Magnetic and Structural Properties
By M. Wuttig, not yet published

132 **Interaction of Hydrogen Isotopes with Transition-Metals and Intermetallic Compounds**
By B. M. Andreev, E. P. Magomedbekov, G.H. Sicking 1996. 72 figs. VIII, 163 pages

133 **Matter at High Densities in Astrophysics**
Compact Stars and the Equation of State
In Honor of Friedrich Hund's 100th Birthday
By H. Riffert, H. Müther, H. Herold, and H. Ruder 1996. 86 figs. XIV, 278 pages

134 **Fermi Surfaces of Low-Dimensional Organic Metals and Superconductors**
By J. Wosnitza 1996. 88 figs. VIII, 172 pages

135 **From Coherent Tunneling to Relaxation**
Dissipative Quantum Dynamics of Interacting Defects
By A. Würger 1996. 51 figs. VIII, 216 pages

136 **Optical Properties of Semiconductor Quantum Dots**
By U. Woggon 1997. 126 figs. VIII, 252 pages

137 **The Mott Metal-Insulator Transition**
Models and Methods
By F. Gebhard 1997. 38 figs. XVI, 322 pages

138 **The Partonic Structure of the Photon**
Photoproduction at the Lepton-Proton Collider HERA
By M. Erdmann 1997. 54 figs. X, 118 pages

139 **Aharonov-Bohm and other Cyclic Phenomena**
By J. Hamilton 1997. 34 figs. X, 186 pages

140 **Exclusive Production of Neutral Vector Mesons at the Electron-Proton Collider HERA**
By J. A. Crittenden 1997. 34 figs. VIII, 108 pages

141 **Disordered Alloys**
Diffusive Scattering and Monte Carlo Simulations
By W. Schweika 1998. 48 figs. X, 126 pages

142 **Phonon Raman Scattering in Semiconductors, Quantum Wells and Superlattices**
Basic Results and Applications
By T. Ruf 1998. 143 figs. VIII, 252 pages

143 **Femtosecond Real-Time Spectroscopy of Small Molecules and Clusters**
By E. Schreiber 1998. 131 figs. XII, 212 pages

144 **New Aspects of Electromagnetic and Acoustic Wave Diffusion**
By POAN Research Group 1998. 31 figs. IX, 117 pages

145 **Handbook of Feynman Path Integrals**
By C. Grosche and F. Steiner. 1998. X, 449 pages

146 **Low-Energy Ion Irradiation of Solid Surfaces**
By H. Gnaser. 1999. 93 figs. VIII, 293 pages

147 **Dispersion, Complex Analysis and Optical Spectroscopy**
By K.-E. Peiponen, E.M. Vartiainen, and T. Asakura. 1999. 46 figs. VIII, 130 pages

148 **X-Ray Scattering from Soft-Matter Thin Films**
Materials Science and Basic Research
By M. Tolan. 1999. 98 figs. IX, 197 pages

Printing: Mercedesdruck, Berlin
Binding: Buchbinderei Lüderitz & Bauer, Berlin